Science Teaching Essentials

Science Teaching Essentials

Short Guides to Good Practice

Cynthia J. Brame
Vanderbilt University, Nashville, TN, United States

ACADEMIC PRESS

An imprint of Elsevier

Academic Press is an imprint of Elsevier
125 London Wall, London EC2Y 5AS, United Kingdom
525 B Street, Suite 1650, San Diego, CA 92101, United States
50 Hampshire Street, 5th Floor, Cambridge, MA 02139, United States
The Boulevard, Langford Lane, Kidlington, Oxford OX5 1GB, United Kingdom

British Library Cataloguing-in-Publication Data
A catalogue record for this book is available from the British Library

Library of Congress Cataloging-in-Publication Data
A catalog record for this book is available from the Library of Congress

ISBN: 978-0-12-814702-3

For Information on all Academic Press publications
visit our website at https://www.elsevier.com/books-and-journals

Working together
to grow libraries in
developing countries

www.elsevier.com • www.bookaid.org

Publisher: Mica Haley
Acquisition Editor: Mary Preap
Editorial Project Manager: Pat Gonzalez
Production Project Manager: Kiruthika Govindaraju
Cover Designer: Matthew Limbert

Typeset by MPS Limited, Chennai, India

Praise for Science Teaching Essentials

"Cynthia J. Brame has produced a very effective, scholarly guide designed to help any teacher seeking to improve how students learn science. Under categories that include Course Design, Assignments and Exams, Active Learning, and Group Work, she concisely presents the latest research conclusions on effective teaching. These evidence-based findings are then connected directly to the classroom through a set of options based on extensive experience. *Science Teaching Essentials* is a wonderful new tool for all those motivated to spread scientific thinking more effectively throughout society."

-Bruce Alberts, PhD, Chancellor's Leadership Chair in Biochemistry and Biophysics for Science and Education, University of California, San Francisco, CA, United States; Former Editor-in-Chief, Science magazine (2008–13); Former United States Science Envoy (2009–11); President Emeritus, US National Academy of Sciences (1993–2005)

"*Science Teaching Essentials* is a much-needed synthesis of the most recent findings in inclusive evidence-based pedagogical practices. Its clear structure is accompanied by relevant and practical examples, making it appealing to any current or future STEM faculty member. It is an excellent reference for professional development experts, and I will certainly recommend it as essential reading to graduate students, postdoctoral scholars, and faculty members participating in evidence-based pedagogy training."

-Laurence Clement, PhD, Academic Career Development Program at the Office of Career and Professional Development; Director of Research in Career Education, University of California, San Francisco, San Francisco, CA, United States

"In *Science Teaching Essentials*, Cynthia J. Brame takes the reader on an enjoyable, exciting journey into the growing world of research analyzing the effects of science teaching practices on student learning. She provides a comprehensive and easily understandable overview of what works, why it works, and how to easily implement best practices into our courses. This will quickly become an essential reference for all instructors who care about quality teaching and making an impact on their students."

-Brent Stockwell, PhD, Professor, Department of Biological Sciences, Department of Chemistry, Columbia University, New York, NY, United States

Contents

Section III
Pedagogy Toolbox

Section IV
Fair and Transparent Grading Practices

Acknowledgments

I want to express my gratitude to several people, without whom I would not have had the time, energy, nor bravery to write this book.

First, I want to acknowledge Derek Bruff, director of Vanderbilt's Center for Teaching. I came to my position at the CFT from a position as a biology faculty member at a small liberal arts college, and Derek was a gracious and wise mentor as I adjusted to my new role. As part of this mentorship, he has recognized the importance of ongoing scholarly work, and he has consistently made it possible for me to have time and space for scholarship. It's ultimately what enabled me to write this book, and I am so grateful.

Second, I want to recognize the myriad contributions of my colleague, Rhett McDaniel, problem-solver, brainstormer, and graphic artist extraordinaire—and, officially, educational technologist at the Vanderbilt Center for Teaching. Rhett was so helpful in helping me think through various ways to talk about, categorize, and present ideas in the book, and, most practically, he created all of the graphics. I am grateful for his help every day.

I've had the privilege to work with an amazing group of graduate students at the CFT in the last several years. I want to thank Ryan Bowen for drafting figures for Chapter 12, Writing Exams: Good Practice for Writing Multiple Choice and Constructed Response Test Questions, and providing feedback on Chapter 1, Inclusive Teaching: Creating a Welcoming, Supportive Classroom Environment. I am so appreciative and look forward to seeing his future work as a science educator. I also want to thank Sara Beck, Justin Hubbard, Kylie Korsnack, and Amie Thurber for their thoughtful commentary on Chapter 11, Incorporating Research Into Courses, and their general encouragement.

I have also had several faculty colleagues who generously took time to read and critique individual chapters. Mary Keithly and Emilianne McCranie Limbrick are former Vanderbilt graduate students who are in their first few years as chemistry faculty members, and they provided valuable commentary on Chapters 1, Inclusive Teaching: Creating a Welcoming, Supportive Classroom Environment, and 2, Course Design: Making Choices About Constructing Your Course. Neuroscientist Anita Disney and biologist Kristy Wilson also provided valuable feedback on Chapter 1, Inclusive Teaching: Creating a Welcoming, Supportive Classroom Environment. I so appreciate the time and thoughtfulness they brought to reading the work.

My undergraduate student assistant, Faith Rovenolt, has been a wonderful partner in bringing this book to fruition. Not only did she coauthor Chapter 11 with me, she also diligently helped with literature searches, reference formatting, and copyediting. I am thankful to work with her!

Finally, I want to thank my husband Andy, reader of early drafts, discusser of ideas and outlines, and general supporter and partner. I can't do it without you.

Chapter Summaries

SECTION I: THE FOUNDATIONS

Chapter 1. Inclusive Teaching: Creating a Welcoming, Supportive Classroom Environment

We all want our classrooms to be inclusive. We want them to be welcoming to all of our students, helping them become part of a science-literate and science-supportive community, and we want them to be productive steps for students moving toward science and science-related careers. This can be challenging; science is hard, our classes are often big, and some of our students arrive with fears that can get in their way. What principles can guide us as we support our students? What practical steps can make our classrooms more inclusive? This chapter defines inclusive classrooms, describes a model explaining the role of a supportive classroom environment in promoting academic achievement, and provides practical steps that instructors can take.

Chapter 2. Course Design: Making Choices About Constructing Your Course

Designing or redesigning a course can be a creative and rewarding effort, but it is always a challenge. Science is characterized by continuous change and an ever-growing (and already large!) body of knowledge, and our courses often seek to help students understand the core knowledge, experimental tools, and ways of thinking in a field. It's a big task. Further, a course may play a particular role in the curriculum, serving as a prerequisite, a capstone, or *the course* in which students learn a particular skill. How do you choose on what to focus, and how do you organize your course to help your students be able to transfer their knowledge to a new setting? How can you design the course to help your students build a conceptual framework that can expand and grow as their understanding grows? This chapter describes six principles to guide your course design and provides suggestions for more detailed resources.

Chapter 3. Assignments and Exams: Tools to Promote Engagement, Learning, and Reflection

What can you have your students do that shows you what they've learned? Meaningful ways to evaluate students' learning are key parts of any course design. Assignments that we grade communicate what we value to students and shape how students engage with the course—and, possibly, how they think of science. We want our graded assignments to serve as tools that allow us to tell whether students have achieved our learning goals and objectives and also serve as positive learning experiences that promote our students' interest and understanding of science. This chapter describes principles that can make assessments more engaging and effective and provides practical suggestions for incorporating these principles in your assignment and exam design.

SECTION II: KEYSTONE TEACHING PRACTICES

Chapter 4. Active Learning: The Student Work That Builds Understanding

Active learning is a cornerstone of the 21st century college science classroom. While it's a bit of a funny term—what learning isn't active?—it has come to mean a collection of teaching approaches that prompt students' active engagement with the course material in a classroom setting. Study after study demonstrates that it's effective at improving student performance, and essay after essay proclaims that it should—or should not—assume a more central role in undergraduate science classes. This chapter explores what active learning is, the role it can play in promoting learning, and some of the evidence that it improves student performance, particularly for under-represented groups. The chapter ends with more than a dozen active learning techniques, from simple to relatively complex, that can be used to supplement or replace parts of lecture.

Chapter 5. Group Work: Using Cooperative Learning Groups Effectively

Group work is the Dr. Jekyll/Mr. Hyde of the college classroom. When it goes well, it can lead to very positive experiences that let students learn from and teach their colleagues as they tackle problems more difficult than they can manage alone. When it goes poorly, it can produce little learning and lots of frustration and resentment. And most of us recognize that there are multiple ways group work can go poorly: students not contributing, students dominating, students arguing within the group, students not engaging with the problem. The challenge is to construct group work to maximize the probability that students will engage and learn. This chapter provides guidance on how to structure group work so that it is a positive and productive experience. It describes the differences between informal and formal group work, evidence that group work can improve students' learning, and instructional choices that can maximize the benefits of these two types of group work.

Chapter 6. Metacognitive Practices: Giving Students Tools to Be Selfdirected Learners

For many scientists, one of our main goals is for our students to become independent, selfdirected learners. Many of the elements discussed in this book can contribute to this goal, from assignments that inspire and engage students to active learning approaches and group work that give them support while they tackle hard challenges. The fundamental key to helping students develop this selfdirection is teaching them to be metacognitive—that is, to monitor and regulate their own learning. This chapter describes definitions for metacognition and reasons that we should foster it in our students. It also suggests specific practices that can be integrated into your course, becoming part of assignments, active learning exercises, group work, and even exams. This regular integration of metacognitive prompts into your course can have a powerful effect on your students' learning, both present and future.

Chapter 7. Test-Enhanced Learning: Using Retrieval Practice to Help Students Learn. With Rachel E. Biel.

Almost all science classes incorporate testing, often as summative assessments at the end of a course segment, formative assessments that measure progress and provide feedback along the way, or diagnostic tools at the beginning of a course. Rarely, however, do we think of tests as learning tools. We may acknowledge that testing promotes student learning, but we often attribute this effect to the studying that students do to prepare for the test. And, yet, one of the most consistent findings in cognitive psychology is that testing is more effective than studying at promoting learning. Given the potential power of testing as a tool to promote learning, we should consider how to incorporate testing into our courses not only to gauge students' learning, but also to promote learning. This chapter defines test-enhanced learning, describes evidence that it can promote learning, and provides suggestions for ways to incorporate testing-for-learning in your course.

SECTION III: PEDAGOGY TOOLBOX

Chapter 8. Lecturing

Lecturing is an essential skill for college science teachers. While we don't want to rely on it as our sole teaching tool, it can serve as a means for instructors to spark students' interest in a new topic, help students develop a framework for a complex subject, and model scientific thinking. Used judiciously and executed well, it can supplement information from readings, frame opportunities for active learning exercises, and help instructors forge relationships with many students at once. The challenge, of course, is structuring our lectures to maximize our students' learning. This chapter describes principles that can be used as a framework for creating effective lectures and provides examples of specific practices that correspond to these principles and enhance students' learning from lecture.

Chapter 9. Flipping the Classroom

The flipped classroom has become an important science pedagogy in which students gain first exposure to new material outside of class and then use class time to do the harder work of assimilating that knowledge. As an educational phenomenon, the flipped classroom has staying power because it relies on two important tools of modern teaching: it can leverage the potential of interactive virtual tools outside of class as well as the known value of active learning approaches in class. The combination has the potential to create powerful learning experiences. This chapter describes the development of the flipped classroom idea, the role it can play in promoting learning, and some of the evidence that improves student performance. The chapter ends with the key elements of the flipped classroom and accompanying practical advice that can help make the flipping experience productive for both faculty and students.

Chapter 10. Using Educational Videos

Video has become an important part of higher education. It is integrated as part of traditional courses, serves as a cornerstone of many blended courses, and is a favored type of resource for students' out-of-class studying. Several meta-analyses have shown that technology can enhance learning and multiple studies have shown that video, specifically, can be a highly effective educational tool. In order for video to achieve its potential as a productive part of a learning experience, however, it is important for instructors to consider several elements of video design and use. This chapter describes a theoretical framework for considering effective design and use of educational videos as well as specific practices that have been shown to enhance students' learning from video.

Chapter 11. Incorporating Research Into Courses. With Faith Rovenolt

There is long-standing and widespread support for educational initiatives that introduce undergraduates to scientific research. Research experiences can lead to gains in students' general skills, such as oral, visual, and written communication, as well as more specific research-associated skills. Traditionally, undergraduates have been introduced to research primarily through the apprentice model, in which undergraduates join a research group and receive one-on-one guidance from a mentor. This model, while effective, limits the number of students who can engage in research experiences and tends to be more accessible to students who are familiar with the research enterprise. There is therefore an increasing emphasis on developing opportunities for students to engage in research in credit-bearing courses, extending the benefits of research experiences to a larger and more diverse group. These course-based research experiences can provide many of the same benefits students derive from traditional undergraduate research experiences, such as increased content knowledge and analytical skills and persistence in science. This chapter describes important characteristics of course-based research experiences and provides examples that range from national programs to single courses developed by individual instructors. We end the chapter by offering practical suggestions to help instructors design an effective course-based research experience.

SECTION IV: FAIR AND TRANSPARENT GRADING PRACTICES

Chapter 12. Writing Exams: Good Practice for Writing Multiple Choice and Constructed Response Test Questions

Giving exams can be one of the most stressful parts of teaching. The challenge begins with the writing: How do you ensure that your exam fairly covers course content? What question constructions help uncover student knowledge? The challenge continues during grading: how do you maximize the probability that exams are graded equitably across the class? And the exams you write also have larger implications: What role do the assessments in your course, including the types of exams students take, play in determining how students study and how they learn? This chapter addresses these questions, first examining important principles to consider when writing exams and then turning to specific recommendations for exam planning and question writing. We end the chapter considering test expectancy, or the role that students' expectations about an exam plays in determining what and how they learn.

Chapter 13. Rubrics: Tools to Make Grading More Fair and Efficient

Science faculty almost universally strive to be objective, fair, and consistent. It's an essential element of our research, and it informs how we approach our classes. We also want our students to have opportunities to do open-ended work that lets them practice and demonstrate more scientific skills and knowledge than we can get at with an exam. Some tension can arise from these two wishes for our classes: how do we help ensure that we are objective, fair, and consistent when we are evaluating work for which there is not a single correct answer? Rubrics provide one tool that can help us align these goals, giving us a means to make our grading more consistent and transparent. They also have the potential to improve students' learning experience and to make our grading more efficient. The key to realizing these benefits is to identify the kinds of rubrics that work for your learning goals, your students, and your assessment style. This chapter briefly explores what rubrics are and the benefits they can offer before providing practical recommendations for how to construct and use rubrics that are a good fit for your needs.

Introduction

Many of us come to teaching with little preparation. We have our own experiences as students to guide us, of course. We may also be able to draw on our experience as research mentors, considering how we have worked with individual students in developing their research projects. These experiences are incredibly valuable and may help us identify key features that we want to incorporate into our teaching, from using questioning to figuring out students' starting points to setting clear goals and expectations. Inevitably, though, these experiences are limited: taking a course as a student doesn't give us access to all the decisions that the instructor makes, and mentoring students one at a time is typically much different than teaching a course with 10 or 30 or 200 students. Many of us, therefore, still feel unprepared to design a course, write a fair and meaningful exam, or figure out how to engage a room full of students for 50 or 75 minutes.

This book intends to help fill that gap, giving scientists clear and practical guidance that they can use to shape their courses and their interactions with students. The book is divided into four sections. The first section focuses on three foundations of effective teaching: principles and practices of inclusive teaching, course design, and assignment and exam development. The second section builds on these foundations by exploring four teaching practices that can be incorporated into any course to enhance students' learning: active learning approaches, group work, metacognitive practices, and test-enhanced learning. These four teaching approaches can be combined and adapted to fit a variety of teaching styles, student populations, and course contexts. They can also be integrated with the four specific pedagogies that are the focus of the third section of the book: lecturing, flipping the classroom, using educational videos, and incorporating research experiences into courses. The final section of the book turns to one of the most challenging elements of teaching: making our assessment and grading fair and transparent, considering both exam construction and rubric design and use. In addition to the chapters that make up these four sections, there are three shorter spotlights that focus on syllabus construction, the use of Bloom's taxonomy for writing learning objectives, and accessible course design.

Although the book has different sections, each with a specific focus, each chapter is intended to be meaningful and useful if read alone. Generally, the chapters open with a description of the topic, explore some underlying principles, and close with practical suggestions that have been shown to be effective in undergraduate science classes. This feature may be particularly important to science faculty: whenever possible, the book includes evidence and citations that the ideas presented have support in college science courses.

In addition to providing evidence and citations for individual practices, there are two models of learning that recur throughout the book and that can be used to situate those individual practices. The first of these is a model for promoting student motivation and development of science identity based on Connell and Wellborn's self-system model for classroom motivation. It can help us think about inclusive teaching practices and effective course and assignment design, as well as why we might use active learning approaches and group work. The second model that appears several times is a model for memory formation based on the Atkinson—Shiffrin model, which can help us think about effective lecture practices and use of video as an educational tool as well as why active learning and retrieval practice help students learn. In both cases, these models are offered not as perfect descriptions of the phenomenon; motivation and memory formation are incredibly complex, and our models don't capture all of the conditional relationships and interactions that influence them. Instead, the models are offered as tools to help us make decisions about our courses within a coherent framework, providing a basis for the ongoing, adaptive process that is teaching.

Thus the goal of this book is to help scientists develop a framework for teaching that incorporates research supported practices but also provides a basis for adaptations and innovations that can promote our students' learning in a particular context or moment. So dive in! Start at the beginning, start in the middle, start at the end; start with the topic that you need the most right now. The goal is for your reading to leave you feeling more prepared for the ongoing and rewarding challenges that teaching brings us.

Section I

The Foundations

Chapter 1

Inclusive Teaching: Creating a Welcoming, Supportive Classroom Environment

We all want our classrooms to be inclusive. We want them to be welcoming to all of our students, helping them become part of a science-literate and science-supportive community, and we want them to be productive steps for students moving toward science-related careers. This can be challenging; science is hard, our classes are often big, and some of our students arrive with fears that can get in their way. What principles can guide us as we support our students? What practical steps can make our classrooms more inclusive? This chapter defines inclusive classrooms, describes a model explaining the role of a supportive classroom environment in promoting academic achievement, and provides practical steps that instructors can take.

WHAT IS AN INCLUSIVE CLASSROOM?

In some ways, it's simplest to consider inclusive environments in contrast with exclusive environments. Exclusive environments by definition exclude some individuals—sometimes intentionally, sometimes not, sometimes based on overt criteria, and sometimes based on unarticulated expectations. In contrast, inclusive environments welcome people with different characteristics, backgrounds, and viewpoints. In some ways, college science classrooms are overtly exclusive: our students must have been admitted to college, and often must have completed prerequisites to enroll in the course. A wealth of research has revealed that our classes have traditionally been exclusive beyond these deliberate criteria, however, and that many voices have consequently been lost from the scientific discourse. This work makes it increasingly clear that we should strive for inclusive classrooms and provides us with tools to do so.

A key element for constructing an inclusive classroom involves the concept of identity. All people within a college classroom have multiple identities, ranging from gender, race, sexuality, and nationality to religious identity, socioeconomic class, and regional identity. The instructor is likely

Science Teaching Essentials. DOI: https://doi.org/10.1016/B978-0-12-814702-3.00001-9

3

to hold "scientist" as an important identity within her list. To make our classrooms inclusive—and thus give our students the opportunity to develop their own identities as "scientist" or "science-friendly" or "science-literate"—we should be attentive to fostering an atmosphere where students can see those science identities as attractive, accessible, and not in conflict with other important ways they see themselves.

On the surface of it, this may seem easy. After all, when we are talking about reading a graph of weather patterns, drawing an evolutionary tree, or determining the terminal velocity of a projectile, we're just talking about data; it does not matter who is doing the work. Our culture is rife, however, with stereotypes that suggest that some people are not good at accomplishing these tasks. These stereotypes can impact both student and instructor behavior, often without their awareness. As humans, we place people into categories based on a variety of characteristics, such as age, gender, race, and role, and we assign expectations—or stereotypes—to these categories (Kang, 2009). Some of our expectations derive from our own experience, but some of them come from things we've read, seen in the media, or heard others say. Unless we are careful to uncover these expectations, they can lead us to respond differently to people in different categories, often unconsciously, such as calling on some students more frequently or being more attentive to their questions. Students are also aware of these societal expectations, and activation of negative stereotypes can decrease performance and aspirations for students in the stereotyped groups. In addition, students may perceive that identities they hold dear are in conflict with course content. In either case—from bias impacting behavior or from perceived conflicts between content and identity—the classroom can become a space where a student's identity interferes with learning. When we strive to create inclusive classrooms, we recognize that potential interference and take steps to stop it.

WHAT CAN MAKE A LEARNING ENVIRONMENT CHILLY OR UNSUPPORTIVE?

Much of what we need to attend to in our classrooms has to do with groups who have experienced overt discrimination such as women, African Americans, and homosexual students. Early research on classroom climate focused on the experience of women in 2- and 4-year college classrooms. Roberta Hall and colleagues identified widespread microinequities for women in the classroom, such as male students receiving more interaction with the professor and more praise and credit for contributions, producing what they referred to as a "chilly" climate for women (Hall and Sandler, 1984; Sandler et al., 1996). Boyer's large survey of undergraduates supported this characterization, finding that faculty called on male students more frequently and took men's contributions more seriously than women's (Boyer, 1987). More recent work suggests that these differences persist

(Eddy et al., 2014; Ferreira, 2003) and extend to other underrepresented groups (e.g., Chang et al., 2011; Brown et al., 2016; Hurtado and Guillermo-Wann, 2013). Further, more than 30 years of research indicate that a supportive climate has a positive influence on students' learning, attitudes, and personal growth and that, in contrast, chilly or hostile environments can impede learning (Freeman et al., 2007; Hurtado and Guillermo-Wann, 2013; Strayhorn, 2012; Whitt et al., 1999; Zumbrunn et al., 2014).

DeSurra and Church expanded the definition of the chilly climate by interviewing approximately 30 gay and lesbian students about their experiences in college classrooms (DeSurra and Church, 1994). They described classrooms as varying along a marginalizing–centralizing continuum, with cues ranging from explicit to implicit. Explicit marginalization stemmed from professors making negative comments about particular groups, while implicit marginalization resulted from professors who avoided issues associated with those groups when they arose within the classroom, such as allowing negative comments by students or in readings to go unchallenged. Alternatively, they described some classrooms where diverse groups were "centralized." In some cases, the instructor planned the course explicitly to include diverse viewpoints or researchers from underrepresented backgrounds. In other cases, the instructors created an "implicitly centralizing," supportive environment by recognizing the importance of underrepresented groups and viewpoints when they were brought to light through student questions or discussion.

We should also strive to be aware that there may be unexpected ways in which classroom cues may signal a conflict with aspects of a student's identity. For example, Rios and colleagues found that Christians and non-Christians recognize the stereotype that Christians are less competent in science, and that this stereotype can reduce Christian students' interest in and identification with science (Rios et al., 2015). Barnes et al. found supporting results in their interviews with Judeo-Christian students, finding that the students reported that their religious identities presented disadvantages in biology classes and could sometimes make the students feel invisible or excluded (Barnes et al., 2017).

HOW DOES CREATING A SUPPORTIVE ENVIRONMENT IMPACT LEARNING?

Carlone and Johnson's model of science identity helps explain why simple actions, such as choosing on whom to call and whose contributions to recognize in class, are important for creating a supportive environment. They identify *competence*, *performance*, and *recognition* as key elements of science identity (Carlone and Johnson, 2007). That is, someone who identifies as a scientist has a deep understanding of science content, can perform scientific practices such as designing, carrying out, and communicating experiments, and is recognized as having these attributes. Of these, recognition by important others—notably science faculty and other science students—is key. Students who feel invisible or unvalued in the classroom are less likely to develop or retain a science identity and to persevere through the challenges associated with pursuing a science degree (Brown et al., 2016).

Research from social psychology has clarified related mechanisms by which classroom climate can impact students. Claude Steele, Joshua Aronson, and others have identified and described stereotype threat as an important factor impacting students' performance. In essence, stereotype threat can occur when a person is a member of a negatively stereotyped group and is taking on a task where that stereotype is relevant, such as a woman taking a math test, an African American taking a science class, or a white man engaging in a test of athletic skill. If the task is challenging and the person cares about performance on it, then signals that remind the person of the stereotype significantly reduces performance (e.g., O'Brien and Crandall, 2003; Steele and Aronson, 1995; Spencer et al., 1999; Stone et al., 1999). These signals are essentially anything that makes people remember their stereotyped identity, from demographic questions on a standardized exam, to casual comments, to gender or racial imbalances (Murphy et al., 2007). This "activation" of the stereotype increases measures of stress and anxiety (Blascovich, et al., 2001), reduces working memory capacity (Schmader and Johns, 2003), and leads to recruitment of neural networks associated with social and emotional processing and reduction in use of neural networks that target the academic task (Krendl et al., 2008), resulting in underperformance on the task. People who are experiencing stereotype threat may be aware that they feel stress and anxiety, but are often unaware of the underlying reasons (Johns et al., 2005). In addition, stereotype threat can reduce aspirations (Davies et al., 2005; Wayne et al., 2010). The effects of activating the stereotype are something of a double whammy, impeding the student's performance and thereby reducing the competence and performance essential for developing a science identity.

A model relating classroom environment, students' sense of belonging, and academic achievement may help clarify the importance of a supportive classroom climate and some specific steps that instructors can take to create it (Fig. 1.1). The model is based on the self-system model of classroom support for motivation described by Connell and Wellborn (1991) and is modified to include results from Zumbrunn and colleagues (2014) and to incorporate the concept of science identity. In essence, the model indicates that a supportive classroom environment that is welcoming for students with various identities produces a sense of belonging. Belonging is essential for the development of self-efficacy and a sense of interest in and value for the course tasks, although particular classroom activities can directly target these elements. Self-efficacy—derived from a student's perception of her competence and ability to perform in the course—and task value, supported by recognition by important others,

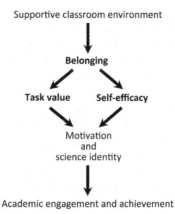

Supportive classroom environment

Belonging

Task value   Self-efficacy

Motivation
and
science identity

Academic engagement and achievement

FIGURE 1.1   Elements that contribute to student motivation and development of science identity. A supportive classroom environment promotes a sense of belonging, which allows students to develop a sense of compentence and value for course tasks; each can promote motivation and a sense of science identity, which lead to engagement with academic activities and academic achievement. Model based on the Connell and Wellborn self-system model for classroom motivation with modifications from Zumbrunn and colleagues (Connell and Wellborn, 1991; Zumbrunn et al., 2014).

produces a science identity (Carlone and Johnson, 2007). The development of these internal elements promotes engagement in academic activities, which then lead to achievement and aspirations.

HOW DO YOU DO IT?

The model described above may be useful in helping instructors think about different avenues they can take to promote an inclusive environment. Here we share some specific strategies for promoting students' sense of belonging, self-efficacy, and value for coursework.

Promoting students' sense of belonging. Perhaps the most important step an instructor can take in helping students achieve is to create an environment where students feel as though they belong and are accepted by their peers and their instructor. This can be accomplished by both generic approaches that apply to all students—such as using student names—and by creating identity−safe environments that explicitly welcome students with different characteristics.

1. *Know students' names or use name tents.*

One way to create a more welcoming classroom is to address students by name, either by learning students' names or by using name tents. Students indicate that having instructors know their name is valuable, reporting that when an instructor knows their names, they feel more valued, more invested in the course, and more likely to seek help (Cooper et al., 2017). Importantly, these benefits derive from students' *perception* that the instructor knows their name, which can be accomplished by the use of name tents in large courses. The use of name tents can also help students build relationships with each other, making it easier for students to form study groups or seek help from peers (Cooper et al., 2017).

2. *Use a welcoming tone in the syllabus.*

A syllabus is often the first point of contact a student makes with a course or instructor, and the tone impacts students' perceptions of how welcoming the course is. Students are more likely to feel comfortable approaching the instructor if the wording of the syllabus uses a promising rather than a punitive voice, such as noting that students submitting late assignments can receive 80% of the total points as opposed to indicating that students who turned in a late assignment will be graded down 20% (Inshiyama and Hartlaub, 2002). Students also rate instructors as more approachable and more motivated to teach the class if they use "warm" language in the syllabus (Hamish and Bridges, 2011). For example, the sentence "I hope you actively participate in this course; I have found it is the best way to engage you in learning the material (and it makes lectures more fun.)" is warmer than "Come prepared actively to participate in this course. This is the best way to engage you in learning the material (and it makes the lectures more interesting.)." Thus, small changes in syllabus wording can impact students' perceptions about the instructor and the probability that they will belong in the class.

3. *Include inclusive language and course policies in the syllabus.*

Including statements in your syllabus that explicitly address students who may have particular needs in your classroom—such as students with disabilities, transgender, or gender-nonconforming students, or students who may be observing religious holidays—indicates that your classroom is an identity–safe environment in which they belong. Incorporating more general statements that indicate that the environment will be one in which differences are respected and valued can help ensure that the welcome extends to all students. Examples can be found on the Vanderbilt Center for Teaching website (cft.vanderbilt.edu) and on the University of Michigan Center for Research on Learning and Teaching website (www.crlt.umich.edu).

4. *Use inclusive language to encourage students to develop leadership skills during group work.* Davies and colleagues worked with first-year medical students in a reproductive physiology course (Wayne et al., 2010). In the control condition, students received the following instructions:

You have 15 minutes to discuss the assigned problem amongst yourselves. Each group will need one volunteer to be group leader to guide discussion of their question. My role as instructor will be to correct misconceptions, answer questions, and provide additional information. The group leader will come to the front of the room, read the question, and then lead the discussion.

In this condition, women were significantly underrepresented among the group leaders (52% women in the class; 33% women group leaders). The instructors then added two sentences to the instructions:

If you've never volunteered to be a group leader in other situations, this is a safe environment to try it out. It doesn't matter what your background is, what your major was as an undergraduate student, whether you're male or female—being a group leader is an important experience for everyone.

The authors found that women's participation as group leaders rose to match their representation in the course. Similar results were observed by Davies et al. (2005).

5. *Include achievements from women scientists and scientists of color.*

Highlighting the work of influential scientists is a common practice in science classrooms, and purposely including the work of underrepresented scientists can help signal that your classroom is identity−safe and thus welcoming of students from a variety of backgrounds. Role models from stereotyped groups have been shown to lead to increased feelings of belonging, greater identification with the subject matter, and improved performance for students in the stereotyped groups (Stout et al., 2011; Young et al., 2013; Carrell et al., 2010). While the greatest impacts are observed for role models with whom students have a significant interaction, some studies have demonstrated that simply reading about successful individuals from different groups can mitigate the effects of stereotype threat and promote belonging (McIntyre et al., 2003). Schinske and colleagues describe using Scientist Spotlight homework assignments to accomplish these goals while also covering content, resulting in changes in student perceptions about who can be a scientist and as well as students' ability to relate personally to scientists, changes that persisted six months after completion of the course (Schinske et al., 2017).

6. *Encourage peer-to-peer interaction and support.*

Belonging in the classroom is as much or more about students' interactions with their peers as their interactions with the instructor (Zumbrunn et al., 2014), and students' sense of community in the classroom correlates with their attitudes and achievement in the class (McKinney et al., 2006). Instructors can promote interaction by including prompts in class for students to work together, such as in a think−pair−share or a clicker-question discussion (for more ideas, see Chapters 4 and 5). By consistently including opportunities for interaction and by supporting collaboration outside of class, instructors can promote the type of community that fosters their students' learning.

7. *Promote a growth mindset.*

Many of our students come to our classes thinking of ability as a fixed entity; they believe they are "good" at a subject or not. Helping students think about intelligence as a malleable entity rather than a fixed entity can increase their sense of belonging, their aspirations, and their academic achievement (Good et al., 2012; Aronson et al., 2002). By teaching students that intelligence is like a muscle, growing with effort, we can help them see themselves as possible participants in the discipline and empower them to develop their ability.

8. *Consider using a social-belonging intervention.*

   Walton and Cohen investigated the impact of a social-belonging exercise on African American and European American first-year college students (Walton and Cohen, 2011). Students were asked to read a narrative that described social adversity in college as an experience that is transient and experienced by most students. The students were then asked to write an essay describing how their college experiences corresponded to this message and to read the essay on video for future college students. African American students who participated in this exercise exhibited improved GPA and health over a three-year period compared to controls; no effect was observed for European Americans.

Promoting students' self-efficacy and sense of task value. Self-efficacy, in essence, is the belief in one's ability to perform a task. It is domain-specific, and has been shown to be an important factor in academic achievement, perseverance, and self-regulated learning (Trujillo and Tanner, 2014). Task value is a description of students' sense that the work of a course is meaningful and valuable to them. It can promote academic achievement by fostering attention, recall, task persistence, and effort (Ainley et al., 2002). By providing students with experiences that help them build belief in their abilities and a belief that the work of the course is valuable, instructors can help students develop a science identity and persist in science. A few suggestions for achieving these goals are described here; others are offered throughout the book, notably in Chapters 3 and 4.

1. *Provide encouragement that emphasizes high expectations and the ability of students to meet them.*

   Social persuasion, or verbal support from instructors or peers, contributes to students' self-efficacy (Trujillo and Tanner, 2014). By acknowledging that a task is challenging but achievable, an instructor can contribute to the students' belief in their ability to accomplish the task. Acknowledging the challenging nature of the task is key; Steele and colleagues have shown that African American students trust and act on feedback framed around high expectations and confidence in the student, but are less likely to respond to feedback that does not include these elements (Cohen et al., 1999).

2. *Provide opportunities for students to practice challenging tasks.*

   One source of self-efficacy is mastery experience, which is also referred to as deliberate practice. In- and out-of-class opportunities that allow students to tackle problems—practicing until they succeed several times—bolsters their self-efficacy with regard to that task, and their subsequent persistence and achievement (Usher and Pajares, 2008; Tanner, 2013). Active learning approaches in class are an effective way to achieve this goal, and have the ancillary benefit of promoting student–student interaction. Multiple active learning approaches are described in Chapter 4.

3. *Make content relevant to students' lives.*

Relating content to real-world problems is a well-supported approach that improves motivation and learning for most students, and it appears to have an enhanced impact for students from underrepresented groups (Hurtado et al., 2007). Many students from underrepresented groups want to pursue science as a way to help their communities, and relating course content to problems that they may see themselves as solving in the future may help them see themselves as belonging to the course and the discipline.

4. *Consider using a values affirmation exercise.*

Miyake and colleagues investigated the impact of a values affirmation activity on women's performance in an introductory college physics class (Miyake et al., 2010). Students in the class were asked to spend 10–15 minutes writing about their most important values and why they were important to them twice during the course, once during class and once on a homework. The exercise was predicted to mitigate the effects of the stereotype that women do not perform as well in physics courses. Women who participated in the values affirmation showed significant improvement in their performance in the course compared to controls, and the activity was particularly beneficial for women who tended to endorse the gender stereotype. The authors suggested that the effects were observed because students' affirmation of their core values helped them reestablish a perception of personal worth. The exercise has been shown to be effective for other stereotyped groups as well.

CONCLUSION

Fostering an inclusive classroom is foundational for our students' learning. It can feel daunting; we do not know all the ways our students differ, and we may feel more or less competent working across various types of diversity. We can, however, consciously take steps to increase students' sense of belonging, create opportunities for students' to develop self-efficacy, and consider ways to make the tasks of the class meaningful and valuable to students. By incorporating these three core principles into our course design and our classroom practice, we signal our intent to create an inclusive environment and to welcome our students—all of our students—to be part of a community that values and understands science. With this as our starting point, we now turn to course design, considering approaches that can help us build courses that effectively promote students' learning about our field.

REFERENCES

Ainley, M., Hidi, S., Berndorff, D., 2002. Interest, learning, and the psychological processes that mediate their relationship. J. Educ. Psychol. 94, 545−561.

Aronson, J., Fried, C.B., Good, C., 2002. Reducing the effects of stereotype threat on African American college students by shaping theories of intelligence. J. Exp. Soc. Psychol. 38, 113−125.

Barnes, M.E., Truong, J.M., Brownell, S.E., 2017. Experiences of Judeo-Christian students in undergraduate biology. CBE Life Sci. Educ. 16, ar15.

Blascovich, J., Spencer, S.J., Quinn, D.M., Steele, C.M., 2001. African Americans and high blood pressure: the role of stereotype threat. Psychol. Sci. 13, 225−229.

Boyer, E.L., 1987. College: The Undergraduate Experience in America. Harper & Row, New York, NY.

Brown, B.A., Henderson, J.B., Gray, S., Donovan, B., Sullivan, S., Patterson, A., et al., 2016. From description to explanation: an empirical exploration of the African-American pipeline problem in STEM. J. Res. Sci. Teach. 53 (1), 146−177.

Carlone, H.B., Johnson, A., 2007. Understanding the science experiences of successful women of color: science identity as an analytic lens. J. Res. Sci. Teach. 44 (8), 1187−1218.

Carrell, S.E., Page, M.E., West, J.E., 2010. Sex and science: how professor gender perpetuates the gender gap. Q. J. Econ. 125 (3), 1101−1144.

Chang, M.J., Eagan, M.K., Lin, M.H., Hurtado, S., 2011. Considering the impact of racial stigmas and science identity: persistence among biomedical and behavioral science aspirants. J. High. Educ. 82, 564−596.

Cohen, G., Steele, C., Ross, L., 1999. The mentors' dilemma: providing critical feedback across the racial divide. Pers. Soc. Psychol. Bull. 25, 1302−1318.

Connell, J.P., Wellborn, J.G., 1991. Competence, autonomy and relatedness: a motivational analysis of self-system processes. In: Gunnar, M., Sroufe, L.A. (Eds.), Minnesota Symposium on Child Psychology: Self-processes and Development. University of Chicago Press, Chicago.

Cooper, K.M., Haney, B., Krieg, A., Brownell, S.E., 2017. What's in a name? The importance of students perceiving that an instructor knows their names in a high-enrollment biology classroom. CBE Life Sci. Educ. 16 (1), ar8.

Davies, P.G., Spencer, S.J., Steele, C.M., 2005. Clearing the air: identity safety moderates the effects of stereotype threat on women's leadership aspirations. J. Pers. Soc. Psychol. 88, 276−287.

DeSurra, C.J. & Church, K.A. November 1994. Unlocking the classroom closet: privileging the marginalized voices of gay/lesbian college students. Paper Presented at the Annual Meeting of the Speech Communication Association, New Orleans, LA.

Eddy, S.L., Brownell, S.E., Wenderoth, M.P., 2014. Gender gaps in achievement and participation in multiple introductory biology classrooms. CBE Life Sci. Educ. 13, 478−492.

Ferreira, M., 2003. Gender issues related to graduate student attrition in two science departments. Int. J. Sci. Educ. 25, 969−989.

Freeman, T.M., Anderman, L.H., Jensen, J.M., 2007. Sense of belongingness of college freshmen at the classroom and campus levels. J. Exp. Educ. 75, 203−220.

Good, C., Rattan, A., Dweck, C.S., 2012. Why do women opt out? Sense of belonging and women's representation in mathematics. J. Pers. Soc. Psychol. 102, 700−717.

Hall, R.M., Sandler, B.R., 1984. Out of the Classroom: A Chilly Climate for Women? Association of American Colleges, Project on the Status and Education of Women, Washington, DC.

Hamish, R.J., Bridges, K.R., 2011. Effect of syllabus tone: students' perceptions of instructor and course. Soc. Psychol. Educ. Int. J. 14, 319–330.

Hurtado, S., Guillermo-Wann, C., 2013. Diverse Learning Environments: Assessing and Creating Conditions for Student Success—Final Report to the Ford Foundation. Higher Education Research Institute, University of California, LA, Los Angeles, CA.

Hurtado, S., Han, J.C., Sáenz, V.B., Espinosa, L.L., Cabrera, N.L., Cerna, O.S., 2007. Predicting transition and adjustment to college: biomedical and behavioral science aspirants' and minority students' first year of college. Res. High. Educ. 48 (7), 841–887.

Inshiyama, J.T., Hartlaub, S., 2002. Does the wording of syllabi affect student course assessment in introductory political science classes? PS: Polit. Sci. Polit. 35, 567–570.

Johns, M., Schmader, T., Martens, A., 2005. Knowing is half the battle: teaching stereotype threat as a means of improving women's math performance. Psychol. Sci. 16, 175–179.

Kang, J., 2009. Implicit Bias: A Primer for Courts. Race & Ethnic Fairness in the Courts, National Center for State Courts.

Krendl, A.C., Richeson, J.A., Kelley, W.M., Heatherton, T.F., 2008. The negative consequences of threat: a functional magnetic resonance imaging investigation of the neural mechanisms underlying women's underperformance in math. Psychol. Sci. 19, 168–175.

McIntyre, R.B., Paulson, R., Lord, C., 2003. Alleviating women's mathematics stereotype threat through salience of group achievements. J. Exp. Soc. Psychol. 39, 83–90.

McKinney, J.P., McKinney, K.G., Franiuk, R., Schweitzer, J., 2006. The college classroom as a community: impact on student attitudes and learning. Coll. Teach. 54 (3), 281–284.

Miyake, A., Kost-Smith, L.E., Finkelstein, N.D., Pollock, S.J., Cohen, G.L., Ito, T.A., 2010. Reducing the gender achievement gap in college science: a classroom study of values affirmation. Science 330, 1234–1237.

Murphy, M.M., Steele, C.M., Gross, J.J., 2007. Signaling threat: cuing social identity threat among women in a math, science, and engineering setting. Psychol. Sci. 18, 879–885.

O'Brien, L.T., Crandall, C.S., 2003. Stereotype threat and arousal: effects on women's math performance. Pers. Soc. Psychol. Bull. 29, 782–789.

Rios, K., Cheng, Z.H., Totton, R.R., Shariff, A.F., 2015. Negative stereotypes cause Christians to underperform in and disidentify with science. Soc. Psychol. Pers. Sci. 6, 959–967.

Sandler, B.R., Silverberg, L.A., Hall, R.M., 1996. The Chilly Classroom Climate: A Guide to Improve the Education of Women. National Association for Women in Education.

Schinske, J.N., Perkins, H., Snyder, A., Wyer, M., 2017. Scientist spotlight homework assignments shift students' stereotypes of scientists and enhance science identity in a diverse introductory science class. CBE Life Sci. Educ. 15, 1–18.

Schmader, T., Johns, M., 2003. Convergent evidence that stereotype threat reduces working memory capacity. J. Pers. Soc. Psychol. 85, 440–452.

Spencer, S.J., Steele, C.M., Quinn, D., 1999. Stereotype threat and women's math performance. J. Exp. Soc. Psychol. 35, 4–28.

Steele, C.M., Aronson, J., 1995. Stereotype threat and the intellectual test performance of African Americans. J. Pers. Soc. Psychol. 69, 797–811.

Stone, J., Lynch, C.I., Sjomeling, M., Darley, J.M., 1999. Stereotype threat effects on Black and White athletic performance. J. Pers. Soc. Psychol. 77, 1213–1227.

Stout, J.G., Dasgupta, N., Hunsinger, M., McManus, M.A., 2011. STEMing the tide: using ingroup experts to inoculate women's self-concept in science, technology, engineering, and mathematics (STEM). J. Pers. Soc. Psychol. 100 (2), 255.

Strayhorn, T.L., 2012. College Students' Sense of Belonging: A Key to Educational Success for All Students. Sage, New York, NY.

Tanner, K., 2013. Structure matters: twenty-one teaching strategies to promote student engagement and cultivate classroom equity. CBE Life Sci. Educ. 12, 322–331.

Trujillo, G., Tanner, K.D., 2014. Considering the role of affect in learning: monitoring students' self-efficacy, sense of belonging, and science identity. CBE Life Sci. Educ. 13 (1), 6–15.

Usher, E.L., Pajares, F., 2008. Sources of self-efficacy in school: critical review of the literature and future directions. Rev. Educ. Res. 78, 751–796.

Walton, G.M., Cohen, G.L., 2011. A brief social-belonging intervention improves academic and health outcomes of minority students. Science 331, 1447–1451.

Wayne, N.L., Vermillion, M., Uijtdehaage, S., 2010. Gender differences in leadership amongst first-year medical students in the small-group setting. Acad. Med. 85 (8), 1276–1281.

Whitt, E., Edison, M.I., Pascarella, E., Terenzini, P.T., 1999. The "chilly climate" for women and cognitive outcomes in the second and third years of college. J. Coll. Stud. Dev. 40, 163–177.

Young, D.M., Rudman, L., Buettner, H., Mclean, M.C., 2013. The influence of female role models on women's implicit science cognitions. Psychol. Women Q. 37, 283–292.

Zumbrunn, S., McKim, C., Buhs, E., Hawley, L.R., 2014. Support, belonging, motivation, and engagement in the college classroom: a mixed method study. Instr. Sci. 42, 661–684.

Chapter 2

Course Design: Making Choices About Constructing Your Course

Designing or redesigning a course can be a creative and rewarding effort, but it is always a challenge. Science is characterized by continuous change and an ever-growing (and already large!) body of knowledge, and our courses often seek to help students understand the core knowledge, experimental tools, and ways of thinking in a field. It's a big task. Further, a course may play a particular role in the curriculum, serving as a prerequisite, a capstone, or *the course* in which students learn a particular skill. How do you pick what to focus on, and how do you organize your course to help your students be able to transfer their knowledge to a new setting? How can you design the course to help your students build a conceptual framework that can expand and grow as their understanding grows? This chapter describes six principles to guide your course design and provides suggestions for more detailed resources (Box 2.1).

BOX 2.1 Course Design: An Overview

Consider the Big Picture
Identify the key learning goals for the course.
Identify one or two big questions that help students see the interest and ongoing importance of the course.
Emphasize the conceptual organization of the course.

Link the Big Picture to Practical Elements
Develop graded assignments that align with your learning goals.
Incorporate formative as well as summative assessments.
Let your learning goals drive your choice of teaching approaches.

WHAT ARE PRINCIPLES TO GUIDE COURSE DESIGN?

Consider the Big Picture

1. *Identify the key learning goals for the course.*

   We often begin thinking about our courses by considering the content we need to cover. It's how our departmental curricula, our textbooks, and

Science Teaching Essentials. DOI: https://doi.org/10.1016/B978-0-12-814702-3.00002-0

even our journals are organized, and it seems natural to start from that perspective. It can be transformative, however, to ask instead what you want your students to be able to do at the end of the course. This question often leads us to think about the big ideas that underpin the course material and how students should be able to use them. For example:

- A genetics course might ask students to be able to describe the mechanisms by which an organism's genome is passed on to the next generation and to predict how the different mechanisms affect the frequency of different types of genetic disorders.
- A physics course might ask students to be able to translate a physical description of a problem to a mathematical equation that can help solve it, and to be able to articulate their expectations for the solution.
- An organic chemistry course might ask students to be able to depict the three-dimensional structure of organic compounds, be able to predict thermodynamically preferred conformations, and to describe how this might affect the rate of a reaction.
- Any science class might ask students to be able to interpret data to draw conclusions and to relate this process to how the field builds knowledge.

In all cases, the course content is essential, but the focus is shifted to consider what the students should be able to do with the content. That shift can help both the instructor and the students. It can help the instructor consider what is really important for students to carry away from the course and what activities can help them reach that goal. It can help students understand not only what they're going to learn, but also why it should matter and what they can do with it. A focus on learning goals, in other words, is a way to explain why what we're doing in the course matters.

Into what categories do learning goals fall? It can be helpful to recognize that learning goals may fall into different categories (Bloom et al., 1956). The most familiar is the cognitive domain, which encompasses the types of intellectual knowledge and skills targeted in the learning goal examples provided above. We may also, however, have goals related to our students' attitudes, motivations, and values. For example, we may want our students to feel empowered to extend their own learning or to value evidence-based claims. Goals like these fall within the affective domain that relates to feelings, values, appreciation, motivation, and attitudes. Finally, we may have goals that require physical movement, coordination, and motor skills, and therefore fall within the psychomotor domain. These skills can range from accurate pipetting, to use of an intricate lab instrument, to specific clinical techniques, and are often critical elements of particular professional progressions. Being explicit about our goals in the affective or psychomotor domain as well as the cognitive domain can help ensure that we build them into the course.

Melanie Cooper and colleagues provide a valuable starting point for identifying learning goals in undergraduate biology, chemistry, and physics courses (Laverty et al., 2016). Using the National Research Council's (2012) *Framework for K-12 Science Education: Practices, Crosscutting Concepts, and Core Ideas* as a starting point, they identify seven scientific practices (Table 2.1) and eight crosscutting concepts (Table 2.2) from the three disciplines as well as core ideas for each that translate to the undergraduate college classroom (Table 2.3). While the work has a solid basis in the NRC's report and was carried out to create a tool for the development of assessment tasks, their adaptations and refinements produce a resource that is particularly valuable for undergraduate instructors considering their learning goals in these disciplines.

How many learning goals should a course have? It is often useful to shape a course around a relatively small number of learning goals that focus on key ideas—perhaps as few as three, perhaps as many as ten. Within the course, however, it is valuable to generate "topic-level learning objectives" that spell out what students should be able to do within different parts of the course at a more granular level. Each of these topic-level learning objectives is associated with a course goal, and can help provide guidance to both the instructor and the student as they think about how

---

TABLE 2.1 Scientific Practices

Asking questions: Generating scientific questions about a real world event, observation, phenomenon, data, scenario, or model.

Developing and using models: Constructing and/or using a mathematical, graphical, computational, symbolic, or pictorial representation to explain or predict an event, observation, or phenomenon.

Planning investigations: Designing an experimental method or identifying a set of observations that can be used to answer a scientific question or test a claim or hypothesis.

Analyzing and interpreting data: Given a question, claim, or hypothesis and relevant data, analyzing the data and interpreting the meaning.

Using mathematics and computational thinking: Using mathematical reasoning or a calculation to interpret an event, observation, or phenomenon.

Constructing explanations and engaging in argument from evidence: Providing reasoning based on evidence to support a claim.

Evaluating information: Making sense of information or ideas.

Source: Adapted from Laverty, J.T., Underwood, S.M., Matz, R.L., Posey, L.A., Carmel, J.H., Caballero, M.D., et al., 2016. Characterizing college science assessments: the three-dimensional learning assessment protocol. PLoS ONE 11(9), e0162333.

TABLE 2.2 Crosscutting Concepts

Patterns

Identifying patterns or trends emerging from three or more events, observations, or data points.

Cause and Effect: Mechanism and Explanation

Identifying mechanistic links between cause and effect.

Scale

Comparing objects, processes, or properties across size, time, or energy scales to identify relevant interactions.

Proportion and Quantity

Predicting the response of one variable to changes in another or identifying the relationship between two or more variables from data.

Systems and System Models

Defining a system, its relevant assumptions and surrounding, and how the system and surroundings interact with each other.

Energy and Matter: Flows, Cycles, and Conservation

Describing the transfer or transformation of energy or matter within or across systems, or between a system and its surroundings, with explicit recognition that energy and/or matter are conserved.

Structure and Function

Predicting or explaining a function or property based on a structure, or describing what structure would lead to a given function or property.

Stability and Change

Determining (1) if a system is stable and providing the evidence for this; or (2) what forces, rates, or processes make a system stable (static, dynamic, or steady state); (3) under what conditions a system remains stable; (4) under what conditions a system is destabilized and the resulting state.

Source: Adapted from Laverty, J.T., Underwood, S.M., Matz, R.L., Posey, L.A., Carmel, J.H., Caballero, M.D., et al., 2016. Characterizing college science assessments: the three-dimensional learning assessment protocol. PLoS ONE 11(9), e0162333.

they will reach the course's broad learning goals during the semester. Examples are provided in Table 2.4, and the Spotlight on Bloom's taxonomy and learning objectives provides more detailed guidance.

Is there evidence that using learning goals helps students learn? Jeffrey Froyd identified the use of learning goals as one of eight promising practices in undergraduate STEM education in a white paper for the National Academies' Board of Science Education (Froyd, 2008). He noted that there is a dearth of studies comparing courses with and without

TABLE 2.3 Core Ideas for Biology, Chemistry, and Physics

Biology Core Ideas	Chemistry Core Ideas	Physics Core Ideas
Chemical and physical basis of life: Life processes are the result of regulated chemical and physical interactions and reactions governed by the laws of physics.	*Electrostatic and bonding interactions:* Attractive and repulsive electrostatic forces govern noncovalent and bonding (covalent and ionic) interactions between atoms and molecules. The strength of these forces depends on the magnitude of the charges involved and the distances between them.	*Interactions can cause changes in motion:* Changes in an object's motion are the result of interactions between it and one or more other objects. Multiple interactions between an object and its surroundings can result in a predictable change in motion.
Matter and energy: Free energy and matter are used in regulated processes that establish order, support growth and development, and control dynamic homeostasis.	*Atomic/molecular structure and properties:* The macroscopic physical and chemical properties of a substance are determined by the three-dimensional structure, the distribution of electron density, and the nature and extent of the noncovalent interactions between particles.	*Energy is conserved:* Energy comes in many forms and can be transformed from one form to another within a given system or transferred between systems.
Cellular basis of life: Cells are the fundamental units of all living things.	*Energy:* Energy changes are either the cause or the consequence of change in chemical systems, which can be considered on different scales and can be accounted for by conservation of the total energy of the system of interest and the surroundings.	*Exchanges of energy increase total entropy:* Multiparticle systems tend toward states that are more statistically likely to occur. At a macroscopic scale, this can be described by concepts such as entropy, temperature, and pressure.
Systems: Ecosystems, organisms, tissues, and cells act as systems.	*Change and stability in chemical systems:* Energy and entropy changes, the rates of competing processes, and the balance between opposing forces govern the fate of chemical systems.	*Interactions are mediated by fields:* Fields are generated by charges/masses. Fields affect charges/masses. In circuits, fields induce currents.
Structure and function: The functions and properties of ecosystems, organisms, tissues, cells, and biological molecules are determined by their structures.		*Energy, momentum, angular momentum, and information can be transported without a net transfer of matter:* Mechanical waves move through matter. Electromagnetic waves can move through vacuum or matter. Properties of waves can be used to parameterize the information or amount of energy, momentum, or angular momentum is transported.
Information flow, exchange, and storage: Hereditary information is stored, used, and replicated.		
Evolution: Evolution drives the diversity and unity of life.		

Source: Adapted from Laverty, J.T., Underwood, S.M., Matz, R.L., Posey, L.A., Carmel, J.H., Caballero, M.D., et al., 2016. Characterizing college science assessments: the three-dimensional learning assessment protocol. PLoS ONE 11(9), e0162333.

TABLE 2.4 Example Course Goals and Learning Objectives

Genetics course goal	*Associated learning objectives*
Students can describe the mechanisms by which an organism's genome is passed on to the next generation and predict how the different mechanisms affect the frequency of different types of genetic disorders.	• Students should be able to diagram the process of meiosis. • Students should be able to explain the events that occur at each stage of meiosis. • Students should be able to compare the predicted frequency of a disorder that arises from a recessive mutant allele and a dominant mutant allele from a given cross.
Organic chemistry course goal	*Associated learning objectives*
Students can interpret data to draw conclusions and can relate this process to how the field builds knowledge.	• Students should be able to identify structural characteristics of small molecules from NMR data. • Students should be able to explain how NMR as a technique allows chemists to understand reaction mechanisms.

learning goals (termed learning outcomes in the paper), but that multiple application studies report their value. Further, the use of learning goals is the centerpiece of most course design frameworks (e.g., Wiggins and McTighe, 2005; Fink, 2003; Chasteen et al., 2011) and is recommended as a key practice for improving undergraduate science education by representatives of the Association of American Universities and the Research Corporation for Science Advancement Cottrell Scholars (Bradforth et al., 2015). Given the value that learning goals can provide for guiding both the students and the instructor, it's a good place to begin course design.

2. *Identify one or two big questions that help students see the interest and ongoing importance of the course.*

Big questions can act as a sort of invitation to a course, an intellectual hook that can draw students into the ways of thinking and the engaging problems within a field. These questions can focus on the foundational ideas and ways of thinking that define science and can be framed around the core concepts in a field, such as those identified in the NRC's *Framework for K-12 Science Education.* For example, the following questions revolve around core concepts in genetics:

- How can individuals of the same species and even siblings have different characteristics?
- How do organisms change over time in response to changes in the environment?
- What evidence shows that different species are related?

- How can there be so many similarities among organisms yet so many different plants, animals, and microorganisms?

These questions are characterized by Wiggins and McTighe (2005) as essential, and can be a valuable way to help students understand both the questions that science seeks to answer and the approaches that are used to answer those questions. Undergraduate science classes can also, however, take things a step further and ask questions that are more explicitly tied to the edge of our understanding, such as:

- Given our growing understanding at each end of the scale of life— from the microbiome to prions—how do we define an organism?
- How do galaxies form and evolve? How do new discoveries challenge our understanding of this process?
- Does our growing understanding of epigenetic events lead to new understanding of "nature vs nurture"?
- What constitutes consciousness?

These questions point to examples where our understanding has changed very recently—and is still changing—and may welcome students to join in the intellectual endeavor of science. They are also big picture questions that encourage students to see connections across courses and disciplines. This combination can be powerful: our students often learn about the process of science in course-associated labs, where they can develop the analytical and experimental design skills but where they may see only a tenuous link to larger questions that shape our understanding of the world. By using one or two provocative, big-picture questions to shape our course, we can capture students' attention, help them understand the potential import of smaller projects, and invite them to be intellectual participants in the scientific journey.

In his study leading to the book *What the Best College Teachers Do*, Bain (2004) found that many transformative college professors design their courses around big questions. They used these questions not only to welcome students to the intellectual work of the course, but also to motivate students to share the instructor's goals. Importantly, these big questions also served as another lens for instructors to consider their learning goals, thinking about the abilities and knowledge students would need to tackle those big-picture questions. Thus questions like these can serve not only as useful hooks to draw your students into the course, but also as a tool to help you define satisfying and meaningful learning goals.

3. *Emphasize the conceptual organization of the course.*

One of the most important factors to consider when designing a course is that your students are relative novices to your field. When faculty members look at a syllabus from their discipline, they do so from the perspective of experts: they have a deep and seemingly intuitive understanding of the key ideas and ways of thinking that underpin the discipline. Students certainly come to the class with knowledge and skills, but they are very unlikely to arrive with a deep-seated understanding of the key features that characterize *biology* or *chemistry* ... much less *genetics* or *physical chemistry*. They therefore encounter the syllabus in a very different way; they see a list of topics but don't yet have a framework to understand how they fit together.

This difference is characteristic of the ways that experts and novices organize their knowledge (Ambrose et al., 2010). Experts have richly connected mental models that allow them to access their knowledge from a variety of points and that allow them to see connections across apparently disparate observations. Novices, on the other hand, tend to have less connected knowledge structures and often have discrete pockets of knowledge or pieces of information that they connect in a linear fashion. In addition, experts also tend to organize knowledge based on deep features rather than surface features, whereas novices often only see the surface features. For example, Kimberly Tanner and colleagues have used a card-sorting task to investigate expert and novice knowledge organization in biology (Bissonette et al., 2017). When given 16 problems, the novices sorted the problems based on the type of organism identified in the problem—a surface feature—while the experts sorted the problems based on key biology concepts, such as structure/function relationships and evolution/natural selection.

By articulating the conceptual organization of the course for yourself, you can emphasize that organization to your students, giving them a more coherent view of the course. This approach has two advantages: it can increase student motivation by helping students see the end goal, and it provides an expert template that students can use to start building their own knowledge structures (Hoskinson et al., 2017).

Link the Big Picture to Practical Elements

4. *Develop graded assignments that align with your learning goals.*

Sometimes, the assessments in our courses are a bit of an afterthought. Many of us hate grading our students—we love our subject, we want students to love our subject, and we see our assessments as unfortunate but necessary interventions that make students study and give us the information we need to assign them a grade. If we make designing important graded assignments an early and integral part of our course design, however, it can change the way we think about assessment, turning it into a tool that helps us refine our learning goals and select learning activities for our students.

In some cases, these assignments may be exams. If so, it can be useful to consider what types of questions will allow students to demonstrate understanding of big-picture concepts, perhaps through working with concrete examples. For example, questions could ask students to identify the underlying problem in a case study and to make recommendations from their analysis of the case; to identify the principle needed to solve a particular problem; to describe a commonly used model and evaluate whether it could be used to predict behavior for a particular example.

In other cases, assignments may take other forms that allow students to demonstrate understanding in less time-constrained settings. For example, students could write a research proposal that illustrates their understanding of a particular topic, identifies an unanswered question, and proposes experiments to answer it. They could design educational materials related to the class for communicating with their peers within the institution, children at local schools, or the general public.

Whatever form the graded assignment takes, the key feature is alignment (Biggs, 2003; Blumberg, 2009). Does it assess how well students are reaching topic-level learning objectives, and does it map back to one of the course's key learning goals? For exams, these questions can be asked on an item-by-item basis; for other types of assessments, alignment may be more holistic. Laverty and colleagues provide tools for evaluating assessments in relation to the scientific practices, crosscutting concepts, and core ideas shown in Tables 2.1−2.3 (Laverty et al., 2016). Aligning assessments with learning goals and objectives increases students' trust in the instructor, places the teaching and learning emphasis on the elements of the course the instructor cares most about, and has the potential to promote student metacognition by giving students a means to compare their self-assessed attainment of learning objectives to the instructor's assessment.

If you are redesigning a course rather than starting from scratch, it may be useful to examine your existing assessments to help you think about your learning goals. Although it's always a good idea for learning goals to drive course design, our existing assessments often have important but unarticulated learning goals embedded in them. Since we often emphasize to students what we consider most important through the points we use, considering existing assignments may help uncover or clarify key things we want our students to learn.

5. *Incorporate formative as well as summative assessments.*

At the beginning of this chapter, we said that we often begin thinking about our courses by considering the content we need to cover. Likewise, when we think about assessment—that is, determining whether students "got" what we want them to get—we often think about exams. Exams and other summative assessments can be important. We often need these relatively high-stakes tools to measure whether students achieved our learning goals and are prepared to move forward in their undergraduate or postgraduate trajectory. It's also important, however, to incorporate formative assessments into the course to allow you and your students to determine how well they are moving toward the learning goals before they need to perform on the summative assessment (NRC, 1999; Couch et al., 2015). Building opportunities for these low- or no-stakes assessments into your course design can help ensure that your students get feedback that can help them succeed in the course.

Formative assessment can take many forms. Some formative assessments can take place in class and be ungraded (or graded for completion). For example, questions that students discuss in small groups before reporting out and hearing the instructor's explanation allow students to evaluate their understanding. If these questions are also answered by all students with clicker-like devices, then they can also let the instructor get a measure of most students' understanding. Cases or problems on which students work in class serve the same function, allowing the students and instructor an opportunity to identify what skills and knowledge students have developed and which areas need more attention. An even simpler form of formative assessment is the "muddiest point" exercise, in which students take 1–3 minutes to write about the idea from the day's class that is most confusing. This serves as a metacognitive exercise for students by prompting them to consider what they do and do not understand and can provide information that can help the instructor determine what additional resources students may need.

Other formative assessments take place out of class. For example, pre-class reading responses that ask students to identify key points and areas of confusion from the reading not only help motivate students to prepare for class but also serve as a metacognitive activity and a point of information for the instructor. Out-of-class problem sets or other homework assignments serve the same purposes, allowing students to practice skills and identify points of confusion and giving instructors a window into students' understanding. In addition, having students submit drafts of papers, posters, or other projects for review can allow identification of areas for improvement—and having some of the review be self- and peer review can both promote student metacognition and keep the instructor's burden manageable.

Angelo and Cross's classic (1993) *Classroom Assessment Techniques* describes many formative assessment approaches with specific instructions for implementation.

6. *Let your learning goals drive your choice of teaching approaches.*

The single most important thing you can do in designing a course—the reason it is listed as the first principle in this chapter and other resources on course design—is to identify your learning goals. Not only do your learning goals serve as a tool for your students to assess their own learning and a guide for you in developing your assessments, they also give you a tool for determining the teaching approaches and learning activities you use.

As noted above, it is useful to identify topic-level learning objectives that tie your course learning goals to specific sections of the course. Those topic-level learning objectives are often powerful ways to think about what your students should do before, during, and after class. Two examples are provided in Table 2.5.

TABLE 2.5 Example Learning Activities Aligned to Course Goals and Learning Objectives

Genetics course goal	Associated learning objectives	Aligned learning activities
Students can describe the mechanisms by which an organism's genome is passed on to the next generation and to predict how the different mechanisms affect the frequency of different types of genetic disorders.	Students should be able to diagram the process of meiosis. Students should be able to explain the events that occur at each stage of meiosis. Students should be able to compare the predicted frequency of a disorder that arises from a recessive mutant allele and a dominant mutant allele from a given cross.	*Before class:* Students read chapter on meiosis and complete short, autograded quiz on your institution's learning management system (e.g., Blackboard or Canvas). *During class:* • Instructor gives mini-lecture on points of confusion identified from quiz and from student questions. • Students diagram meiosis in pairs and write short explanations of events. • Instructor chooses one example diagram to project; class reviews, corrects the example diagram and their own diagram as necessary. • Instructor poses a problem: What would be the outcome of a mutation in a protein necessary for crossing over? Students discuss in pairs and vote for one of four responses. • Instructor poses a follow-up question: Would your prediction differ for a dominant mutation vs. a recessive mutation? Students discuss in pairs and vote. • Instructor requests student explanations of their responses and then uses those to provide correct response. *After class:* Students complete additional problems asking them to predict frequency of disorders from particular crosses and, conversely, asking them to interpret the crosses that could result in particular outcomes.

(Continued)

TABLE 2.5 (Continued)

Organic chemistry course goal	Associated learning objectives	Aligned learning activities
Students can interpret data to draw conclusions and can relate this process to how the field builds knowledge.	Students should be able to identify structural characteristics of small molecules from NMR data. Students should be able to explain how NMR as a technique allows chemists to understand reaction mechanisms.	*Before class:* Students watch videos describing how to interpret NMR (e.g., https://my.vanderbilt.edu/ochem2/homework-2/), completing a short worksheet to help them remember key elements. *During class:* • Instructor gives a mini-lecture summarizing key points and responding to student questions. • Students complete worksheet in small groups assessing their ability to apply the concepts from the videos. Example: https://my.vanderbilt.edu/ochem2/files/2015/01/Workshop1.pdf. The instructor circulates, answering individual questions as needed and stopping the whole class at major points of confusion. *After class:* Students do a homework problem in which a reaction and several possible products are shown. The NMR spectrum of the observed product is provided. Students interpret the NMR spectrum to identify the product and explain how the spectrum allows them to differentiate among the possible products.

In both examples, the learning activities are explicitly tied to the topic-level learning objectives. They use independent student work, small group work, and instructor explanation to help students develop the ability to meet the learning objectives. Further, there are opportunities for the students and the instructor to assess student understanding, because formative assessment and learning activities are often intrinsically linked. Finally, after completing these learning activities, students should be able to achieve the learning objectives—or should have a clear sense that they cannot and that they should seek additional help and practice prior to the summative assessment.

The examples provided in the table are not the only learning activities that could help students reach these learning objectives, of course. For example, instead of having students watch videos before class, the instructor could intersperse lecture with short opportunities for students to practice—essentially alternating between the information that is in the videos and in the example worksheets in the second example. What is important, however, is that the learning activities are clearly tied to the learning objectives: students know what they are supposed to be able to do and are given opportunities to practice those skills with peer interaction and instructor feedback.

CONCLUSION

An essential part of designing a course is determining for yourself and sharing with your students why the course matters. What are students going to learn to do? What big questions will they consider? What will they come to understand about the way your discipline organizes knowledge? Design that considers these big-picture questions results in inviting, exciting courses, giving students a sense that the course will be a compelling and rewarding experience. Of course, the second essential piece of course design is linking the big picture to practical elements. Developing assignments that align with (and help students fulfill) your learning goals, giving students chances to practice and get feedback in a low-stakes way, and choosing teaching approaches that are a good match for your learning goals are critical for fulfilling the big picture promise.

There are several well-established frameworks for course design that can provide more detailed guidance for developing a course. Two of the most widely used are Grant Wiggins' and Jay McTighe's "backwards design" process, described in *Understanding by Design*, and Dee Fink's integrated course design process described in *Creating Significant Learning Experiences: An Integrated Approach to Designing College Courses*.

Thus far, we have considered inclusive approaches to teaching and principles of effective course design, two foundational elements for science teaching. We now turn to approaches to developing assignments and exams, considering principles that ensure that these critical assessments are an integrated part of our courses.

REFERENCES

Ambrose, S.A., Bridges, M.W., DiPietro, M., Lovett, M.C., Norman, M.K., 2010. How Learning Works: Seven Research-Based Principles for Smart Teaching. Jossey-Bass, San Francisco, CA.

Angelo, T.A., Cross, K.P., 1993. Classroom Assessment Techniques: A Handbook for College Teachers, second ed. Jossey-Bass, San Francisco, CA.

Bain, K., 2004. What the Best College Teachers Do. Harvard University Press, Cambridge, MA.

Biggs, J., 2003. Aligning teaching and assessing to course objectives. Teaching and Learning in Higher Education: New Trends and Innovations. University of Averio, pp. 13–17.

Bissonnette, S.A., Combs, E.D., Nagami, P.H., Byers, V., Fernandez, J., Le, D., et al., 2017. Using the biology card sorting task to measure changes in conceptual expertise during postsecondary biology education. CBE Life Sci. Educ. 16 (1), ar14.

Bloom, B.S., Krathwohl, D.R., Masia, B.B., 1956. Taxonomy of Educational Objectives: The Classification of Educational Goals. D. McKay, New York, NY.

Blumberg, P., 2009. Maximizing learning through course alignment and experience with different types of knowledge. Innov. High. Educ. 34, 93–103.

Bradforth, S.E., Miller, E.R., Dichtel, W.R., Leibovich, A.K., Feig, A.L., Martin, J.D., et al., 2015. University learning: improve undergraduate science education. Nature 523, 282–284.

Chasteen, S.V., Perkins, K.K., Beale, P.D., Pollock, S.J., Wieman, C.E., 2011. A thoughtful approach to instruction: course transformation for the rest of us. J. Coll. Sci. Teach. 40, 70–76.

Couch, B.A., Brown, T.L., Schelpat, T.J., Graham, M.J., Knight, J.K., 2015. Scientific teaching: defining a taxonomy of observable practices. CBE Life Sci. Educ. 14, 1–12.

Fink, D.L., 2003. Creating Significant Learning Experiences: An Integrated Approach to Designing College Courses. Jossey-Bass, San Francisco, CA.

Froyd, J.E., 2008. White paper on promising practices in undergraduate STEM education. Commissioned paper for the Evidence on Promising Practices in Undergraduate Science, Technology, Engineering, and Mathematics (STEM) Education Project, The National Academies Board on Science Education.

Hoskinson, A.-M., Maher, J.M., Bekkering, C., Ebert-May, D., 2017. A problem-sorting task detects changes in undergraduate biological expertise over a single semester. CBE Life Sci. Educ. 16 (2), ar21.

Laverty, J.T., Underwood, S.M., Matz, R.L., Posey, L.A., Carmel, J.H., Caballero, M.D., et al., 2016. Characterizing college science assessments: the three-dimensional learning assessment protocol. PLoS ONE 11 (9), e0162333.

National Resource Council, 1999. In: Bransford, J.D., Brown, A.L., Cocking, R.R. (Eds.), How People Learn: Brain, Mind, Experience, and School. National Academies Press, Washington, DC.

National Research Council, 2012. A Framework for K-12 Science Education. National Academies Press.

Wiggins, G., McTighe, J., 2005. Understanding by Design. Association for Supervision and Curriculum Development, Alexandria, VA.

Writing Learning Objectives Using Bloom's Taxonomy

Writing learning objectives is one of the most powerful steps you can take to maximize the educational experience for both you and your students. The language around learning objectives can be complicated, with the lines among learning goals, objectives, and outcomes blurry and indistinct, but the concept is simple. Learning objectives put the endpoint of a lesson front and center, helping both instructors and students sharpen their focus. They help instructors identify what they want their students to be able to do, helping them determine what to assess, what to do in class, and what to have students practice. They allow students to understand better what it means to "know" something in the class, helping them move beyond memorization and toward other skills that allow them to use the facts they've learned. In short, learning objectives are a tool to marry the content of a class to the ways of thinking and communicating that characterize our disciplines; they are a means to convert a class from a list of topics into a series of lessons on how to think scientifically about those topics.

Topic-level learning objectives can serve as sort of roadmap for both students and instructors, illustrating what students will do to reach the large learning goals that are shaping the course. Table S1 gives an example of three sets of topic-level learning objectives by which students in a statistics course build toward the overarching course goals.

Learning objectives have two key parts: a verb that describes what student should be able to do, and a description of the content they'll be working with. Bloom's taxonomy, described by Bloom et al. (1956) and revised by Anderson et al., (2001), can serve as a valuable tool in helping instructors think through their choices for both parts of the learning objective.

Bloom's taxonomy describes three types of learning (cognitive, affective, and psychomotor) and four knowledge categories. Considering Bloom's description of cognitive processes can help instructors consider what they want students to be able to do—the verb in the learning objective—and considering the knowledge categories can clarify the type of content they want to be the focus of students' work.

TABLE S1 Course-Level Learning Goals May Be Reached Through Multiple Sets of Learning Objectives During a Course's Trajectory

Course-Level Learning Goal	Associated Learning Objectives: Set 1	Associated Learning Objectives: Set 2	Associated Learning Objectives: Set 3
Students will be able to use statistics to analyze the results of a study.	Students will be able to choose the appropriate Student's t-test for a comparison of two populations, identifying key assumptions.	Students will be able to choose the appropriate statistical test to compare the effect of a single independent variable on more than two populations, identifying key assumptions.	Students will be able to choose the appropriate statistical test to compare the effect of two variables on more than two populations, identifying key assumptions.
	Students will be able to apply the appropriate Student's t-test to a given dataset.	Students will be able to apply the chosen test to a given dataset.	Students will be able to apply the chosen test to a given dataset.
	Students will be able to interpret and explain the results of the t-test.	Students will be able to interpret and explain the results of the test.	Students will be able to interpret and explain the results of the test.

The example shown here focuses on statistical methods.

COGNITIVE PROCESSES: WHAT DO YOU WANT YOUR STUDENTS TO BE DOING?

Within the cognitive learning domain, Bloom's revised taxonomy describes six categories of cognitive processes, building from simple to complex:

- *Remembering:* Retrieving knowledge from memory through recognizing or recalling information.
- *Understanding:* Constructing meaning. Can be demonstrated by interpreting and paraphrasing, generating examples, classifying, summarizing, inferring, comparing, or explaining.
- *Applying:* Executing or implementing a procedure.

Can the student...

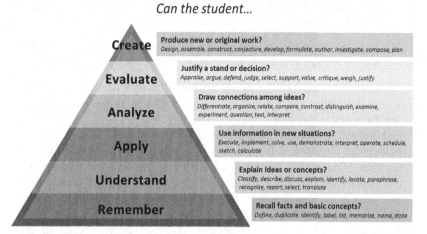

FIGURE S1 **Bloom's revised taxonomy describes six levels of cognitive processes.** Figure from Vanderbilt University Center for Teaching.

- *Analyzing:* Determining how parts of a whole relate to each other and to an overall structure or purpose by differentiating, organizing, or attributing.
- *Evaluating:* Making judgements based on criteria and standards by checking or critiquing.
- *Creating:* Synthesizing elements into a coherent whole, such as generating hypotheses or research plans.

The categories are summarized in Fig. S1, where they are shown in the triangle that we often associate with Bloom's taxonomy. Generally, remembering and understanding are considered lower-order cognitive processes, whereas applying, analyzing, evaluating, and creating (or synthesizing) are considered higher order. While we want our students to remember and understand content from our discipline, we typically also want them to be able to do more, and these categories can help us recognize the forms that the "more" can take.

Instructors often find Bloom's cognitive processes to be helpful particularly when the basic six categories are illustrated by a rich set of verbs such as those shown in Fig. S1; a quick google of "Bloom's verbs" will lead to even more extensive lists.

When writing learning objectives, it's critical to recognize that not every level will be represented for every topic and that higher level cognitive activities are not necessarily better. Bloom's categories are useful tools for thinking about the learning we want our students to do, but it's important for instructors to use their understanding of the topic and the discipline as a whole to identify goals for their students.

THE KNOWLEDGE DOMAIN: WHAT TYPES OF KNOWLEDGE DO WE WANT OUR STUDENTS TO LEARN?

Blooms' taxonomy is most well-known for characterizing cognitive processes, but the revised taxonomy also describes four knowledge categories.

- *Factual:* The basic pieces of information students need to be familiar with a discipline and the approaches it uses to solve problems, such as key facts and terminology.
- *Conceptual*: Understanding relationships between basic pieces of factual knowledge to create a more complex knowledge structure that functions together. Conceptual knowledge includes knowledge of classifications and categories; principles and generalizations that describe those categories; and theories and models that a discipline uses to describe, explain, and predict phenomena.
- *Procedural*: Procedural knowledge involves an understanding of how processes work (e.g., skills, algorithms, techniques, and methods, and the criteria for using them).
- *Metacognitive:* Knowledge of different strategies for learning, their relative effectiveness, and conditions under which they might be used, as well as an understanding of personal knowledge and learning processes.

These knowledge categories can be used to prompt our thinking about the kinds of things we want our students to know. We certainly want them to know factual information from our classes, but we also want them to organize this factual organization in a conceptual framework and to understand the processes we use to ask and answer questions in our discipline. These distinctions are often not apparent to students. Experts tend to organize their knowledge in richly connected networks based on essential features, whereas novices tend to have more fragmented knowledge networks and tend to base their networks on surface features (Ambrose et al., 2010). If we are not careful, therefore, our students will have a tendency to memorize a great number of facts from our classes but not necessarily organize them into meaningful conceptual networks. By using Bloom's knowledge categories to help us write our learning objectives, we can be intentional about prioritizing the types of conceptual understanding that can help our students develop transferable knowledge. Table S2 illustrates learning objectives that vary both across the type of thinking students are doing and the type of knowledge that is their focus.

In brief, writing learning objectives is one of the most powerful steps you can take to maximize the educational experience for both you and your students. Bloom's taxonomy can serve as a tool to help instructors identify the cognitive processes they want their students to be able to do—the verb in the learning objective—as well as the type of content they want to be the focus of students' work.

TABLE S2 Learning Objectives Can Vary in Both Level of Cognitive Process and Type of Knowledge

Cognitive Process	Topic: Michaelis–Menten Kinetics	Topic: Mechanisms of Enzyme Regulation	Knowledge Category
	Associated Learning Objectives Could Be:	Associated Learning Objectives Could Be:	
Remember	Students should be able to write the Michaelis–Menten equation and a typical Michaelis–Menten plot.	Students should be able to cite examples of enzymes regulated by posttranslational modification, allosteric regulators, and genetic control.	Factual
Understand	Students should be able to explain the meaning of the variables in the Michaelis–Menten equation and conditions under which it applies.	Students should be able to explain how different types of enzyme regulation lead to changes in enzyme activity.	Conceptual
Apply	Students should be able to produce the Michaelis–Menten plot from experimental data.	Students should be able to predict the impact of a particular inhibitor on enzyme K_m and V_{max}.	Procedural
Analyze	Students should be able to interpret changes in reaction conditions from different Michaelis–Menten plots.	Students should be able to analyze the impact of a given mutation on regulation of that enzyme.	Conceptual
Evaluate	Students should be able to determine whether the Michaelis–Menten relationship can be used to identify parameters for a particular enzyme, justifying the reasons.	Given an unfamiliar metabolic pathway and the ΔG for each enzyme, students should be able to select the most likely major point of control for the pathway, justifying their answers.	Conceptual
Create	Students should be able to design an experiment to generate data for a Michaelis–Menten plot.	Students should be able to design an experiment to test the role of allosteric regulation on the activity of a given enzyme.	Conceptual and procedural

The two examples shown illustrate multiple levels of understanding for two topics from a biochemistry class.

REFERENCES

Ambrose, S.A., Bridges, M.W., DiPietro, M., Lovett, M.C., Norman, M.K., 2010. How Learning Works: Seven Research-Based Principles for Smart Teaching. Jossey-Bass, San Francisco, CA.

Anderson, L.W., Krathwohl, D.R., Bloom, B.S., 2001. *A Taxonomy for Learning, Teaching, and Assessing: A Revision of Bloom's Taxonomy of Educational Objectives* (Complete ed.). Longman, New York, NY.

Bloom, B.S., Krathwohl, D.R., Masia, B.B., 1956. Taxonomy of Educational Objectives: The Classification of Educational Goals. D. McKay, New York, NY.

Chapter 3

Assignments and Exams: Tools to Promote Engagement, Learning, and Reflection

What can you have your students do that shows you what they've learned? Meaningful ways to evaluate students' learning are key parts of any course design. Assignments that we grade communicate what we value to students and shape how students engage with the course—and, possibly, how they think of science. We want our graded assignments to serve as tools that allow us to tell whether students have achieved our learning goals and objectives but that also serve as positive learning experiences that promote our students' interest and understanding of science. This chapter describes principles that can make assessments more engaging and effective and provides practical suggestions for incorporating these principles in your assignment and exam design.

WHAT ARE THE PRINCIPLES TO GUIDE DEVELOPMENT OF ASSIGNMENTS AND EXAMS?

In considering the principles that can help us design assignments and exams, it may be useful to reconsider the model of motivation and science identity development introduced in Chapter 1 (see Fig. 3.1). This model suggests that a supportive classroom environment is necessary for students to develop a sense of belonging. This sense of belonging helps students develop a sense of value for the tasks of the course (task value) and a belief in their ability to complete these tasks (self-efficacy). Collectively, belonging, task value, and self-efficacy lead to student motivation and development of a science identity, both of which promote academic engagement and success.

When considering assignments and exams, two elements of this model are particularly relevant: task value and self-efficacy. If students believe that their assignments and exams are valuable and that they can do them well, they will be more motivated and learn more from the experiences.

Promoting task value. A primary driver of student engagement is interest in a task or topic. Perhaps unsurprisingly, interest promotes academic

Science Teaching Essentials. DOI: https://doi.org/10.1016/B978-0-12-814702-3.00003-2

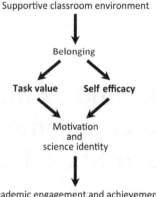

Supportive classroom environment

Belonging

Task value   Self efficacy

Motivation
and
science identity

Academic engagement and achievement

FIGURE 3.1   Elements that contribute to student motivation and development of science identity. When considering assignments and exams, the task value and self-efficacy elements of this model are particularly relevant; assignments that are engaging and that students believe they can accomplish promote motivation and development of science identity. Model based on the Connell and Wellborn self-system model for classroom motivation with modifications from Zumbrunn and colleagues (Connell and Wellborn, 1991; Zumbrunn et al., 2014).

achievement, fostering attention, recall, task persistence, and effort (Ainley et al., 2002). It also shapes the paths our students choose to follow. Short-term, or situational, interest in one element of a class—including a challenging, meaningful assignment or an engaging exam—can lead to an enduring interest that encourages students to take more science courses or even to pursue a science career (Harackiewicz and Hulleman, 2010). Thus one of the most important things we can do when developing an exam or assignment is to design it with student interest in mind.

The value students place on a task can arise from their intrinsic interest or their perception that the task has usefulness to their lives. One of the easiest ways to maximize task value is to provide students with an element of choice in our assignments; we can give students an opportunity to act on their intrinsic interest or to adapt an assignment such that it has some utility for their lives. It's key, of course, to ensure that the assignment still aligns with the learning objectives, allowing the students an opportunity to demonstrate the appropriate skills and knowledge. Providing students some autonomy in the exact ways they reach those learning objectives can be powerful, however.

Of course, like us, our students have many competing demands on their lives, and providing student choice may lead some of our students to make the easiest choice. For this reason, it may be particularly important within an assignment to make an explicit link to students' lives or to ask students to identify such a link. For example, Harackiewicz and Hulleman (2010) found that asking students to consider how a single topic in a psychology course applied to their lives increased student interest in the field as a whole at the end of the semester, particularly among students who had a lower expectation of success. This result supported previous work showing similar results in high school science students (Hulleman and Harackiewicz, 2009). It is particularly notable because interest is also linked to expectations for success, with students who expect to do well demonstrating greater persistence, performance, and interest in academic work; in this study, increasing task value overcame some of the effect of lower expectations.

Promoting self-efficacy. When we ask our students to tackle interesting tasks that demonstrate that they're doing the kinds of higher order thinking we've identified as important in our learning objectives, we are often asking them to take on challenging work that they may not know how to organize or structure. Coupled with other tasks competing for their time, this can lead students to put off starting the work or to allot less time than it requires, leading to disappointing products. Further, they may not recognize the kinds of work they need to be doing to accomplish the task or even what a really good product would look like. To help students tackle these complex tasks and understand the level of work they need to do, we can introduce scaffolding that helps them pull apart the project into manageable chunks, enabling them to better show their ability to reach the goals for the project and simultaneously helping them develop a toolkit for managing future projects.

Scaffolding is, in essence, structuring assignments to make the goals and process clear to students (Skene and Fedko, 2017). Scaffolding can be structured to help students understand the process of completing a complex project by breaking it down into smaller components and providing feedback as the project develops. Alternatively, scaffolding can be structured to help students progressively develop the types of cognitive skills they need, moving from lower order assignments that ask students to summarize or describe, to assignments where they analyze and evaluate, to culminating assignments where they build on the skills they have developed earlier to create a new approach. Finally, scaffolding can focus on evaluation, helping students consider what the goals of an assignment are and how a final product should look. Further, evaluation scaffolding can provide tools for students to reflect on their work, promoting their metacognitive activity (a topic explored in more detail in Chapter 6).

HOW DO YOU DO IT?

Strategies for assignments. Enhancing our assignments relies on increasing students' perception of the assignment's value and increasing students' belief that they can effectively complete the task. The first suggestion provided here is an approach that can increase students' sense of task value, and the remaining suggestions offer mechanisms to increase students' self-efficacy through process scaffolding, cognitive skills scaffolding, or evaluation scaffolding. The options offered can overlap in powerful ways and can be considered a menu of sorts to optimize assignments.

1. *Consider the "students-as-producers" model.*

    Giving students an element of choice is a powerful way to enhance interest in an assignment, and one way to do that is to use a model in which students generate a product that has value outside the classroom. In our college courses, we often think of students as consumers of knowledge, learning the key concepts and skills they need to move on in their chosen fields. In "students-as-producers" assignments, however, the focus is shifted to students acting as producers of knowledge as we do in our professional lives. The name originated at the University of Lincoln (http://studentasproducer.lincoln.ac.uk/) and reflects the type of thinking that led to the Freshman Research Initiative at the University of Texas at Austin (Rodenbusch et al., 2016) and other efforts to get college students in the lab early. It can also, however, be used to guide the development of assignments that provide a direct avenue to promote student interest and investment. Derek Bruff has helped faculty in multiple disciplines adopt this model (see e.g., https://prezi.com/ggrxifv0g-kz/students-as-producers-northwestern/) and has identified three key elements to a students-as-producers assignment (Bruff, 2013):

    - *It deals with an open-ended problem.* Students-as-producers projects don't have a single best, known answer; like our research, they deal with open-ended problems where either we don't know the answer or there are multiple possible answers.

    - *It has an authentic audience.* Successful students-as-producers projects are shared with an audience beyond the instructor. Students may share their work with the other students in the class or at the institution, with a particular local audience, or with the larger academic world, but the key is that the work has an authentic audience that goes beyond the bounds of the student–instructor circle, increasing the relevance of the work.

    - *There is a degree of autonomy.* Giving students some choice and control in shaping their project helps maximize their engagement. While the instructor is responsible for ensuring that the assignment is aligned with course goals and objectives, assignments that have some flexibility and allow space for student autonomy can be powerful.

    What can these assignments look like? In their course on health policy at Vanderbilt University, Gilbert Gonzalez and Tara McKay asked students to choose a health policy topic related to a proposed bill in the state legislature. Working in groups of three or four, students developed informative and attractive two-page briefs that they shared and discussed with legislators on a visit to the state capital. It was a powerful learning experience for the students as well as a boon for under-resourced state legislators. In her course on Drugs and Behavior, neuroscientist Anita Disney gave her students two objectives for their final assignment—but then left the shape of the project they completed open for their design. Some students designed and implemented education plans; others wrote a series of poems illustrating a class of drugs along with an artist's statement; others wrote more traditional research papers. She had students working on similar types of projects—for example, creative or educational projects—work together to develop rubrics to make grading feasible and fair, thereby providing another opportunity for student control. The key in both of these cases was giving students a degree of autonomy and the opportunity to develop meaningful projects.

2. *Process scaffolding: creating markers to help students move toward the final product.*

When confronted with a relatively large project in your class, your students may need help identifying the steps they need to take and the best order for proceeding. It can therefore be helpful for the instructor to consider the order in which students ought to complete the assignment to do their best work and structure mini-assignments to help that happen. Table 3.1 provides an example of process scaffolding for a lab report. This example suggests that students should begin writing their report by thinking about the experiment they did and the results they found, expressing these results in figures and tables before moving on to add descriptive and framing text. The introduction and abstract are left to later in the process, after the student has a sense about what the results say. Process scaffolding for an argumentative essay might ask students to begin by identifying a topic, critiquing relevant papers, and then outlining an argument before writing significant amounts of text. In both cases, this type of ordering is not intuitive; students will often start writing the introduction before taking the steps they need to prepare.

It can also be valuable to turn some of the work of scaffolding over to your students, enhancing their sense of control. As students embark on a larger project that will culminate in a paper, a presentation, a poster, or another product worth significant value in your class, have them write a plan for how they will break the project into parts and complete it. When will they do each step, and what resources will they need? How will they leave time for revision and, if desired, getting feedback from others? Asking students to do this kind of planning as part of their classwork often feels uncomfortable to science instructors; we don't like to be micromanaged ourselves and are reluctant to do it to others. Our students, however, often don't yet have the project management skills that are developed through graduate school and professional life, and don't recognize the need to carve out time and to plan ahead for the various steps of a project; giving them a brief, graded-for-completion assignment that encourages them to do that can help them become aware of the

TABLE 3.1 Scaffolding Examples

Process Scaffolding: Lab Report	Cognitive Skills Scaffolding: Literature Critique	Evaluation Scaffolding: Presentation
• Student writes experimental methods section, describing work that was done. • Student creates figures and tables with accompanying explanatory captions/legends. • Student writes results text, using figures and tables as guideposts. • Student writes discussion and introduction. • Student writes abstract and title. • Student puts together elements of the report to produce a final product.	• Student reads abstract and introduction of two conflicting articles and summarizes the arguments for the class. • Student analyzes experimental design and results in the two papers, drawing conclusions and identifying limitations. • Student identifies reasons for conflict in two papers. • Student proposes an experiment to clarify point of conflict. • Student writes final paper, synthesizing the elements above to make a cohesive argument about the next steps needed to clarify understanding of the scientific question.	• Student works with a small group to identify key elements of an effective presentation. • Instructor facilitates a whole-class discussion to construct of a rubric based on small group decisions. • Students generate checklist for self-evaluation based on rubric.

importance of this step. (Having them reflect on how well they followed the plan after the project can be an additional beneficial step.)

3. *Cognitive skills scaffolding: helping students develop the types of thinking they need for the project.*

We do not always recognize the complexity of the tasks we ask students to complete, and thinking through the elements of producing a good critique or a good argument can point us to the steps we need to help our students take to get there. Table 3.1 provides an example of cognitive skills scaffolding for a literature critique, describing how a student moves from summarizing to analysis and critique to synthesis. In addition, the Backwards Faded Scaffolding process used by Stephanie Slater and Tim Slater provides a great example of scaffolding that both helps students develop cognitive skills and an understanding of a scientific process, focusing specifically on inquiry skills in astronomy labs (Slater et al., 2010). Students begin by drawing conclusions from provided data before moving on to collecting their own data from which to draw conclusions. They then develop the research procedure, collect data, and draw conclusions to answer a predetermined question. Finally, students ask their own question that can be answered through a scientific approach. Because asking the question is the most challenging task students do, it is the one where they get support the longest—the one where the scaffolding fades last.

4. *Evaluation scaffolding: helping students think about the quality of their work.*

One of the best ways to help students do well on challenging assignments is to incorporate scaffolding that helps them think about the quality of their work. Annotated examples are one effective way to do this. By providing a couple of examples of student work and highlighting key elements that were done well, as well as elements that should be improved, the instructor can help students see how the abstract ideas in the assignment description play out in practice.

Students can also build their own understanding of the expectations for an assignment by working together to design a rubric. If you are assigning a larger project—a writing assignment, an oral presentation, a poster, etc.—asking students to spend part or all of a class period designing a rubric can help ensure that they have thought about the level of work they should be producing and that they are invested in the grading criteria. Specifically, you can have students work in small groups to identify characteristics that should be examined in grading the project and then identify markers of excellent, good, and mediocre performance for those characteristics.

We can also encourage our students to use our rubrics—generated by students or by the instructor—as tools for self-reflection (Reddy, 2007). If students complete a draft of their project at least a week before the final project is due, they can use a rubric for self-evaluation and note areas for improvement, and then submit the final project with these notes, helping them become more reflective about the quality of their work. This process not only provides the students with a great opportunity to consider what they're learning and how to improve, it also gives the instructor additional information about the students' process, such as how deeply they're thinking about the project and how well they follow through on plans for revision. This approach will be most effective if it's completed in class, where students have time set aside for the sometimes painful process of self-evaluation as well as access to the instructor's guidance. In addition, it can be helpful to talk to the students about the purpose of the process to ensure that they see it as valuable rather than busywork.

Peer review can also help students consider the quality of their own work—and can remove some of the burden of providing feedback from the instructor. Carefully structured peer review processes, where students raise questions and provide suggestions about their colleagues' projects, help the reviewers to think about their own projects, and also provide students with an audience beyond the instructor. There are tools for download (e.g., Calibrated Peer Review, http://cpr.molsci.ucla.edu/Home.aspx) and purchase (e.g., Peerceptiv, https://go.peerceptiv.com/) that can facilitate peer review in larger classes, but the essential process is relatively simple and familiar to most instructors—and can be a powerful tool for helping scaffold students' work on complex projects. While peer review can be helpful to students, receiving feedback from the instructor on elements of their work is invaluable. It can provide scaffolding but also can make a student feel recognized and help establish an important academic relationship.

Strategies for exams. Enhancing exams to promote student motivation and development of a science identity also relies on promoting students' perception that the exams are valuable and that the students can do well. The mechanisms by which we do this differ from the mechanisms used for assignments, however. In particular, the types of scaffolding that we can use don't fall neatly into the categories described above. The first suggestion provided here is an approach that can increase students' sense of an exam's value, and the remaining suggestions offer scaffolding mechanisms that can increase students' self-efficacy.

1. *Consider case studies or context-rich problems.*

   Exam questions that are linked to cases or context-rich problems (see Box 3.1 for an example from physics) can increase students' ability to see how course material has real-world applicability and usefulness for their lives. These questions tend to engage students by providing an opportunity for them to apply their knowledge. In addition, they can give students a sense of how they may use the knowledge from the course in the future, providing a prompt for seeing science as a useful tool in their current and future lives. Cases that can be used as inspiration for exam questions can be found on CourseSource (http://www.coursesource.org/) or the National Center for Case Study Teaching in Science (sciencecases.lib.buffalo.edu).

   Problems like these do have higher cognitive load than simple, context-free problems, however. To take advantage of the benefits of context-rich problems while managing the cognitive load they impose, it may be helpful to adopt one or more of the following strategies:
   - Clearly delineate and separate the questions students should answer about the case. This allows students to address each relevant question and reduces the cognitive load required to monitor whether they've included all of the information they should to answer a larger, more open-ended question.
   - Allow use of a "cheat sheet." Context-rich questions and cases are intended to elicit higher order thinking, but students often also have

BOX 3.1 Example Context-Rich Problem From a Physics Exam

Used with permission from Shane Hutson, Department of Physics & Astronomy, Vanderbilt University.

One of your patients tore her Achilles tendon, which connects the calf muscles to the heel of the foot. She has been rehabbing the injury for 6 months and is scheduled to visit your office later today. She wants to know when she can start running again, so you need to evaluate how much tension her Achilles tendon can now comfortably bear. Your plan is to have her attempt to stand on her tip-toes using just her injured leg—first just lifting her body weight, then lifting her body weight while you also push down on her shoulders (with a force of roughly 70% of *your* body weight). You will need to give your patient an assessment of how much tension her Achilles can bear under these conditions, and compare this results to the Achilles tensions she will experience when she starts running again (up to 6—7 times *her* body weight).

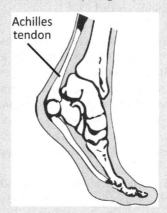

Achilles tendon

Your patient is 5'6" tall and weighs 130 lbs. You pull up an X-ray of her foot and measure a few distances. It's 25 cm from the ball of her foot to the point where her Achilles tendon attaches to her heel, and 19 cm from the ball of her foot to the center of her ankle joint (where the weight-bearing tibia bone of her lower leg meets the bones of her foot). You then look up a few additional pieces of information that you think you might need: a typical adult female foot weighs about 1% of body weight; the center-of-mass of the foot is roughly halfway between the heel and the ball of the foot; and when standing on one's tip-toes, the Achilles tendon is aligned vertically when the foot makes an angle of 25° with respect to the floor.

Experience tells you that your patient will readily accept that she can't start running again if she can't even pass the first assessment—that is, standing on the tip-toes of her injured leg when you aren't pushing down on her shoulders—but you also know that you'll need to justify why she can't start running if she passes the first assessment, but fails at the second (when you push down on her shoulders). To make sure you're ready to provide this justification, write down a

(Continued)

> **BOX 3.1 (Continued)**
>
> *quantitative argument* detailing what you learn from each assessment and why she must pass the second one before she can safely start running again. This argument should include the calculation of a relevant quantity or quantities, their comparison in the two assessments, and their comparison to what she should expect when she starts running again.

to remember key equations or definitions to demonstrate that thinking. Allowing the use of a cheat sheet—generated by students during study or by the instructor to equalize available content—can eliminate anxiety associated with remembering those facts and focus students on the type of thinking needed.

- Allow some part of the exam to be done in groups. For example, Meyer begins physics exams with a group discussion phase, where students are given a "number-free" version of a context-rich problem (Meyer, 2016). Students work together to digest the story-like problem and explore solution strategies before receiving the full problem and beginning the individual part of the exam. Knierim and colleagues also found benefits for collaborative sections on exams in a geology course, finding that individual student performance increased in a course in which exams had a collaborative component (Knierim et al., 2015).

- Allow the case to be done as a take-home problem. Allowing students to work at their own pace, without the time constraints associated with an in-class exam, reduces one source of anxiety and thus one source of cognitive load. If the problem or case is one that requires independent thinking and can't readily be found online, allowing it to be done out of class can help students construct better, more nuanced answers that give the instructor a better sense of their understanding.

2. *Scaffolding.*

Scaffolding for exams occurs primarily before the exams, as instructors help students develop the skills they need to do well. In addition to the standard work that students are doing—reading, homework, in class lectures, and exercises—instructors may also incorporate specific activities to help students prepare for exams.

- One way we can help students prepare for an exam, and thus develop a sense of confidence about how they will perform, is to teach them about specific study strategies that are effective. Students often spend most of their study time reviewing their notes and their text, often with a highlighter in heavy use, without realizing that these are some of the least effective strategies for learning (Dunlosky et al., 2013).

Retrieval practice, in which students practice recalling information on a given topic, and self-explanation, in which they say or write explanations as they answer questions, have repeatedly been shown to be more effective at promoting learning. As students recall and explain their understanding, they are strengthening retrieval pathways and forging links to existing knowledge, creating interconnected knowledge structures that improve student understanding and performance. Chapter 7 describes more about the evidence that retrieval practice promotes learning and suggests specific strategies for incorporating it into courses.

- Another way to help students prepare for an exam is to teach strategies for self-assessment. One approach is to use a knowledge survey. Students examine a previous exam, using a Likert scale to indicate their confidence that they could answer the question (Yu et al., 2008). After completing the survey, students can then complete the exam questions and compare to a key, identifying areas where their assessment corresponded with their performance and areas where it did not and adjusting their studying to fill the gaps.
- Other instructors have students develop and critique exam questions, using the instructor's goals and Bloom's taxonomy of cognitive skills to guide their work. Discussing those questions within the class can allow an assessment of how well the questions fit the instructor's goals and can provide students a useful study tool. This type of activity can be particularly helpful in ensuring that students are aware of the level and type of work they'll need to do on the exam.
- After an exam, it can be useful to ask students to reflect on their performance. There are several ways to accomplish this. Exam wrappers are short reflections that students complete in class after an exam, typically describing how much time they spent preparing, how they prepared, where they ran into trouble on the exam (e.g., trouble applying definitions; careless mistakes), and what they plan to do differently for the next exam (see e.g.: https://www.cmu.edu/teaching/designteach/teach/examwrappers/). In some cases, instructors take up the reflections, review them, and return them a few weeks before the next exam. The wrappers provide a powerful way to get students to look beyond the number at the top of the exam page, to consider what elements of the exam with which they struggled, and to connect their study behavior with that performance. Alternatively, Mynlieff and colleagues observed that students who completed written corrections to exam questions showed significant improvement on subsequent assessments (Mynlieff et al., 2014). In both cases, the reflection on exam performance can encourage students to identify areas of strength and areas for improvement, helping them feel prepared to tackle the next exam.

CONCLUSION

Assessments—both exams and project-like assignments—can be designed to promote student interest and value for the task and can be scaffolded to enhance students' skill development and belief in their ability to complete the task effectively. When we take the trouble to design our assessments in this way, we transform them from being only a means to grade students into powerful components of our courses that also promote students' learning.

In this first section of the book, we have considered three foundations for effective teaching: developing an inclusive, supportive classroom; taking an intentional approach to course design; and designing assessments that promote student interest and motivation. The next section of the book builds on these foundations, considering four adaptable teaching and learning practices that can enhance students' experiences in any course.

REFERENCES

Ainley, M., Hidi, S., Berndorff, D., 2002. Interest, learning, and the psychological processes that mediate their relationship. J. Educ. Psychol. 94, 545–561.

Bruff, D. 2013. <https://cft.vanderbilt.edu/2013/09/students-as-producers-an-introduction/> (accessed 07.09.17).

Connell, J.P., Wellborn, J.G., 1991. Competence, autonomy and relatedness: a motivational analysis of self-system processes. In: Gunnar, M., Sroufe, L.A. (Eds.), Minnesota Symposium on Child Psychology: Self-processes and Development. University of Chicago Press, Chicago, IL.

Dunlosky, J., Rawson, K.A., Marsh, E.J., Nathan, M.J., & Willingham, D.T., 2013. Improving students' learning with effective learning techniques: promising directions from cognitive and educational psychology. Psychol. Sci. Public Interest 14, 4–58.

Harackiewicz, J.M., Hulleman, C.S., 2010. The importance of interest: the role of achievement goals and task values in promoting the development of interest. Soc. Pers. Psychol. Comp. 4, 42–52.

Hulleman, C.S., Harackiewicz, J.M., 2009. Promoting interest and performance in high school science classes. Science 326, 1410–1412.

Knierim, K., Turner, H., Davis, R.K., 2015. Two-stage exams improve student learning in an introductory geology course: logistics, attendance, and grades. J. Geosci. Educ. 63, 157–164.

Meyer, C., 2016. Group work tests for context-rich problems. Phys. Teach. 54 (5), 302–305.

Mynlieff, M., Manogaran, A.L., St. Maurice, M., Eddinger, T.J., 2014. Writing assignments with a metacogntive component enhance learning in a large introductory biology course. CBE Life Sci. Educ. 13 (2), 311–321.

Reddy, Y.M., 2007. Effects of rubrics on enhancement of student learning. Educate 7 (1), 3–17.

Rodenbusch, S.E., Hernandez, P.R., Simmons, S.L., Dolan, E.L., 2016. Early engagement in course-based research increases graduation rates and completion of science, engineering, and mathematics degrees. CBE Life Sci. Educ. 15 (2), pii: ar20.

Skene, A., Fedko, S., 2017. Scaffolding Assignments. Centre for Teaching and Learning, University of Toronto, Retrieved from: <http://ctl.utsc.utoronto.ca/home/sites/default/files/scaffolding.pdf> (accessed 07.09.17).

Slater, S.J., Slater, T.F., Lyons, D.J., 2010. Engaging in Astronomical Inquiry. W.H. Freeman Publishing Company, New York.

Yu, S., Wenk, L., & Ludwig, M. (2008). Knowledge surveys. In: Session Presented at the National Association of Geoscience Teachers (NAGT) Workshops: The Role of Metacognition in Teaching Geosciences, Carleton College, Northfield, MN. <http://serc.carleton.edu/NAGTWorkshops/metcognition/tactics/28927.html> (accessed 07.09.17).

Zumbrunn, S., McKim, C., Buhs, E., Hawley, L.R., 2014. Support, belonging, motivation, and engagement in the college classroom: a mixed method study. Instr. Sci. 42, 661−684.

Spotlight 2

Considerations for Syllabus Writing

Your course syllabus has several key functions (Gunert O'Brien et al., 2008). It establishes an early point of contact between the student and instructor, helping to set the tone for the course and conveying information about the course's purpose and design. It also provides logistical information, from required materials and course policies to the types of assessments in the course. Importantly, it can also convey a welcome to all your students by including information about learning resources, statements about inclusivity, and statements on accommodation. This insert describes syllabus elements that instructors may want to consider, including example statements and introduction activities.

Tone. Many of us think about the syllabus as a contract with students, and we therefore have a tendency to use impersonal language that sets out policies and expectations. This can result in a syllabus that is clear and informative but that students perceive as cold and distant. In *What the Best College Teachers Do*, Ken Bain found that highly effective instructors invite their students to the course, emphasizing the excitement and the shared journey on which they'll go. Students perceive these differences, and they associate "warm" language with the qualities of a master teacher (Saville et al., 2010) who is approachable and motivated to teach the class (Hamish and Bridges, 2011). It's therefore worth reading your syllabus with an eye to its tone. An easy way to warm up the language is to convert from third person to first and second person. For example, converting the sentence "Students will design a final project that ties together course objectives and future career goals" to "You will have the opportunity to design a final project that ties the course to your future career goals" conveys the same information but in a warmer and more inviting way. Further, converting the sentence "Students must actively participate throughout the course" to "Your voice matters. We will have frequent opportunities for you to contribute to the class, and it's important for you to take advantage of them to enhance your own and your colleagues' learning" personalizes the message and provides a rationale that centers on the student. These changes can be a powerful way to communicate our goals about the type of course and the types of relationships we want to develop.

Expectations for students/expectations for instructor. The syllabus often describes course expectations for students through attendance policies, class preparation policies, phone/laptop use policies, etc. Some faculty find it valuable to have a list of expectations for students with a mirroring list of expectations for the instructor. For example, neuroscientist Anita Disney includes the following section in her syllabus:

You can expect me to:

- *Respect each of you as individual and unique learners.*
- *Be on time and prepared for each class.*
- *Turn off my mobile phone.*
- *Make instruction clear and offer further clarification when help is needed.*
- *Respond to emails within 24 hours.*

Here is what I expect from you:

- *Respect your classmates and support them in their endeavors.*
- *Be on time for class.*
- *Complete your assigned readings.*
- *Silence or turn off your phone; make no calls, send no texts.*
- *Provide credit and citation where it is due—in class, online, and in your submitted assignments.*

The reciprocal nature of these statements demonstrates respect for students and can have a positive impact on the course community.

Inclusivity and diversity statements. A statement about your or your institution's support for inclusive learning environments can be a way to welcome students to your class and to signal your support for diverse student groups. Vanderbilt's Center for Teaching (cft.vanderbilt.edu) has collected several statements that can serve as inspiration (Riviere et al., 2016); one from the University of Northern Colorado is shown here:

- *The College of Education and Behavioral Sciences (CEBS) supports an inclusive learning environment where diversity and individual differences are understood, respected, appreciated, and recognized as a source of strength. We expect that students, faculty, administrators and staff within CEBS will respect differences and demonstrate diligence in understanding how other peoples' perspectives, behaviors, and worldviews may be different from their own.*

Disability. Although it falls under the general category of inclusivity, it may also be worthwhile to have a section on your syllabus that directly addresses disability. Students in our classes may have a range of abilities and disabilities; some of these differences will be visible and some will not. A statement in the syllabus that invites students with disabilities to meet with you privately to discuss accommodations is a good start to making the classroom welcoming. Picard and colleagues offer the following example statement (Riviere et al., 2016):

- *This class respects and welcomes students of all backgrounds, identities, and abilities. If there are circumstances that make our learning environment and activities difficult, if you have medical information that you need to share with me, or if you need specific arrangements in case the building needs to be evacuated, please let me know. I am committed to creating an effective learning environment for all students, but I can only do so if you discuss your needs with me as early as possible. I promise to maintain the confidentiality of these discussions. If appropriate, also contact the Equal Opportunity, Affirmative Action, and Disability Services Department (EAD) to get more information about specific accommodations.*

Personal electronic devices. Student use of cell phones, tablets, or laptops can enrich a class, providing a mechanism to find resources, answer clicker questions, or take notes directly on the images we often use to guide our class. Student misuse of devices, on the other hand, can serve as distractions for the entire class. It's worthwhile considering how you expect students to use electronic devices in class, develop a corresponding policy, and explain it in the syllabus. It's important to avoid adversarial language when introducing the policy and to keep your explanation grounded in students' learning experiences. Riviere and colleagues offer the following statements (Riviere et al., 2016):

- *As part of this course, you will be asked to vote on "clicker questions" using your laptop, tablet, or cell phone at least twice per class. To ensure that you gain as much as possible from class time, I expect that you use these technologies only for course-related purposes while in the classroom. If you find your cellphone distracting, I encourage you to put it away between questions.*
- *This course requires you to be a mindful and courteous participant during in-class discussions. Therefore, laptops and cell phones are not allowed except in the following situations: fact-checking, referencing required readings, and finding relevant resources to aid in our understanding of the course content (such as YouTube clips or recent newspaper articles). If I find that you are not following this policy, I will ask you to turn off the device.*

Activities to get students to read the syllabus. After you've done the work of putting together a warm syllabus that invites students to your exciting course, it can be discouraging when they don't seem to read it. Rather than "going over the syllabus" with your students on the first day of class, it can be useful to put together an activity that gets students actively engaged in finding key information. The following are two examples of activities that can be effective.

- *Syllabus mining.* Have your students form groups of three and provide different questions for each person in the group, such as:

    Person 1: How are synthesis maps made? How do you get feedback on them, and when are they graded? When are they due, and we will do anything in class to help with their construction?

    Person 2: What are the recommendations for reading papers? What do you need to do to lead discussion? How will your discussion leadership be graded? Are your peers involved in grading you? Where do you find information about dates for each paper discussion?

    Person 3: What are the recommendations for reading the text? Will we use Brightspace? Will we use any other tool for communication? When are the two exams?

    Students in the small group share their answers with each other. You then ask the large group if anything needs clarification or additional information. This exercise not only gets students actively involved in finding key information in the syllabus, it also gets them talking to and relying on each other in the first class period.
- *Syllabus speed dating.* Have students sit in two rows of chairs facing each other. Ask two questions, one about something in the syllabus and one of a more personal nature. Give a brief amount of time for students to answer both question, check that the syllabus question has been answered correctly, and then have one row of students shift down to form new pairs (Weimer, 2017).

In brief, your syllabus is one of your first chances to establish the way your course will proceed, from the tone of your interactions with students to the things you value in your class. Using it to welcome students, convey the course's purpose and design, and share important logistical information makes it a multi-functional document that will be useful to you and your students, and designing activities that ensure that students engage with it ensures that they get the most out of it.

REFERENCES

Grunert O'Brien, J., Millis, B.J., Cohen, M.W., 2008. The Course Syllabus: A Learning-Centered Approach, second ed. Jossey-Bass, San Francisco, CA.

Hamish, R.J., Bridges, K.R., 2011. Effect of syllabus tone: students' perceptions of instructor and course. Soc. Psychol. Educ. 14, 319–330.

Riviere, J., Picard, D. R., Coble, R., 2016. Syllabus design guide. Retrieved from: <http://cft.vanderbilt.edu/guides-sub-pages/syllabus-design/> (accessed 26.04.18).

Saville, B.K., Zinn, T.E., Brown, A.R., Marchuk, K.A., 2010. Syllabus detail and students' perceptions of teacher effectiveness. Teach. Psychol. 37 (3), 186–189.

Weimer, M., 2017. First day of class activities that create a climate for learning. Faculty Focus. <https://www.facultyfocus.com/articles/teaching-professor-blog/first-day-of-class-activities-that-create-a-climate-for-learning/> (accessed 19.07.17).

Spotlight 3

Making Our Courses Accessible: Universal Design for Learning

As members of the scientific community think more about how to make classes inclusive and welcoming, there is a growing interest in developing courses that are accessible for students with varying learning strengths and needs. Universal Design for Learning (UDL) is one framework that can help instructors develop course materials, activities, and assessments that are adaptable to student needs. UDL is based on the architectural movement of Universal Design, which established principles for making spaces usable by people with varying abilities without the need for accommodation. UDL embraces a similar approach for teaching, suggesting that courses provide multiple means for student engagement, content representation, and student contributions to the class.

CAST describes a set of guidelines that breaks these large categories into smaller chunks (CAST, 2018). For example, the guidelines suggest that students' engagement is enhanced when instructors provide options for prompting student interest, sustaining students' efforts, and fostering students' self-regulation. These goals can be met by a variety of approaches that are good teaching practice for all of our students: emphasizing the relevance and value of the class, providing students some autonomy in choosing course elements, fostering collaboration and community, providing goal-directed feedback, and promoting students' metacognition. Similarly, varying content representation involves effective teaching practices such as activating or supplying background knowledge and highlighting patterns, critical features, and relationships, as well as offering alternatives for auditory and visual representation. Finally, providing options for student contributions can involve using multiple media for instruction and assessment.

There are a couple of core ideas within these guidelines that have the potential to allow all of our students to do their best work. First, if we provide students with multiple ways to demonstrate their understanding, we are more likely to be able to make an authentic judgement about their understanding, both to provide feedback and to produce a grade. Some instructors accomplish this goal by incorporating multiple types of assignments during a semester, from exams, papers, and presentations to problem sets and podcasts, reasoning that all students will have a particular affinity for one or more of the assignments. Other instructors provide choice within a major class project, giving students learning objectives that they must accomplish but leaving the specific topic and the format of project—essay, video, poster—up to the student. Maryellen Weimer describes an alternative in which she combines these two approaches, giving students a menu of assignments (each with learning objectives and a strict due date) from which they can choose to construct their suite of activities in the course. Whatever specific form this core idea takes in your context, all of our students benefit when we provide them a chance to demonstrate their knowledge beyond a standard, timed exam.

A second core idea that cuts across the UDL guidelines involves providing multiple ways for students to communicate within the course. We all have students who love to talk in class, offering their responses to every question we raise. We also, however, have students who are much more introverted and hesitant to speak out. By offering multiple ways that students can contribute to course conversations, we foster community and learning for all of our students. There are several easy ways to do this:

- Use the think—pair—share approach to soliciting student answers. Pose a question, ask students to discuss it with their neighbor for a specified length of time, and then request a volunteer to share their group's answer and reasoning. Many students are more able to discuss a question with a partner than in a large group setting.
- Modify the think—pair—share to the think—write—share. Instead of having students discuss their response with a colleague, have them instead write a response. Quieter students who need a moment to collect their thoughts will be more inclined to respond after writing.
- Use clickers (or other personal response devices). Pose a multiple choice question, ask students to vote on the answer, show the graph depicting student responses. Ask for volunteers to share their reasoning. This approach gives all students a chance to commit to an answer and receive feedback, and can also be used to prompt small or large group discussion.
- Use the minute paper. Pose a question, have students spend 1−5 minutes writing a response, and then collect the responses. In large classes, this approach can work best at the end of class, while in smaller classes, it can be used throughout. Because the instructor collects the responses, all students have a chance to contribute in a way that can inform how the class proceeds.
- Use online discussion boards and blogs. Some students are more comfortable sharing their questions and comments electronically, so it can be useful to supplement in-class activities with opportunities to contribute online.

- Consider holding online office hours. Many learning management systems offer a tool for online office hours. Zoom and Skype are alternatives not associated with an LMS. Not only do online office hours offer students an electronic buffer that can be helpful for students with anxiety, they also provide few accessibility barriers for students who have problems with mobility.

In short, the UDL framework can help instructors make choices that enhance accessibility of the course and that align with teaching practices that promote learning for all students. It can be a powerful tool to help instructors think through their options for reaching their goals.

REFERENCE

CAST, 2018. Universal design for learning guidelines version 2.2. Retrieved from: <http://udl-guidelines.cast.org>.

Section II

Keystone Teaching Practices

Chapter 4

Active Learning: The Student Work That Builds Understanding

Active learning is a cornerstone of the 21st century college science class-room. While it's a bit of a funny term—what learning isn't active?—it has come to mean a collection of teaching approaches that prompt students' active engagement with the course material in a classroom setting. Study after study demonstrates that it's effective at improving student performance, and essay after essay proclaims that it should—or should not—assume a more central role in undergraduate science classes. This chapter explores what active learning is, the role it can play in promoting learning, and some of the evidence that it improves student performance, particularly for under-represented groups. The chapter ends with more than a dozen active learning techniques, from simple to relatively complex, that can be used to supplement or replace parts of lecture.

WHAT IS IT? A WORKING DEFINITION FOR ACTIVE LEARNING

There are various definitions of active learning, but they all revolve around the idea that students use their knowledge to tackle a problem, and in doing so, test and extend their understanding. To help expand this definition and illustrate variations, Box 4.1 provides definitions from several important sources.

Bonwell and Eison's seminal work on active learning defined it as "instructional activities involving students in doing things and thinking about what they are doing" (Bonwell and Eison, 1991). They emphasize that approaches that promote active learning focus more on developing students' skills than on transmitting information and require that students do something—read, discuss, write—that requires higher-order thinking.

This definition is broad, and Bonwell and Eison explicitly recognize that a range of activities can fall within it. They suggest a spectrum of activities to promote active learning, ranging from very simple (e.g., pausing lecture to

Science Teaching Essentials. DOI: https://doi.org/10.1016/B978-0-12-814702-3.00004-4

BOX 4.1 Active Learning Definitions

Instructional activities involving students in doing things and thinking about what they are doing.

Bonwell and Eison (1991)

Active learning implies that students are engaged in their own learning. Active teaching strategies have students do something other than taking notes or following directions . . . they participate in activities . . . [to] construct new knowledge and build new scientific skills.

Handelsman et al. (2007)

Active learning engages students in the process of learning through activities and/or discussion in class, as opposed to passively listening to an expert. It emphasizes higher-order thinking and often involves group work.

Freeman et al. (2014)

Students' efforts to actively construct their knowledge.

Carr et al. (2015)

allow students to clarify and organize their ideas by discussing with neighbors) to more complex (e.g., using case studies as a focal point for decision making). In their book *Scientific Teaching*, Handelsman, Miller, and Pfund note that the line between active learning and formative assessment is blurry and hard to define; after all, teaching that promotes students' active learning asks students to do or produce something, which then can serve to help assess understanding (Handelsman et al., 2007).

The National Survey of Student Engagement (NSSE) and the Australasian Survey of Student Engagement (AUSSE) provide a very simple definition: active learning involves "students' efforts to actively construct their knowledge." This definition is supplemented by the items that AUSSE uses to measure active learning: working with other students on projects during class; making a presentation; asking questions or contributing to discussions; participating in a community-based project as part of a course; working with other students outside of class on assignments; discussing ideas from a course with others outside of class; tutoring peers (reported in Carr et al., 2015).

Freeman and colleagues collected written definitions of active learning from >300 people attending seminars on active learning, arriving at a consensus definition that emphasizes students' use of higher-order thinking to complete activities or participate in discussion in class (Freeman et al., 2014). Their definition also notes the frequent link between active learning and working in groups.

Thus, active learning is commonly defined as activities that students do to construct knowledge and understanding. The activities vary but require students to do higher-order thinking. Although not always explicitly noted, metacognition—students' thinking about their own learning—is an important element, providing the link between activity and learning. While some definitions stipulate that these activities occur during class, it's clear that students can engage in active learning processes outside of class as well.

WHAT'S THE THEORETICAL BASIS? OR, WHY SHOULD IT WORK?

Constructivist learning theory emphasizes that individuals learn through building their own knowledge, connecting new ideas and experiences to existing knowledge and experiences to form new or enhanced understanding (Bransford et al., 1999). The theory, developed by Piaget and others, posits that learners can either assimilate new information into an existing framework, or can modify that framework to accommodate new information that contradicts prior understanding. Approaches that promote active learning often explicitly ask students to make connections between new information and their current mental models, extending their understanding. In other cases, teachers may design learning activities that allow students to confront misconceptions, helping students reconstruct their mental models based on more accurate understanding. In either case, approaches that promote active learning promote the kind of cognitive work identified as necessary for learning by constructivist learning theory.

Taking a slightly different slant, we can also use a model of memory formation (used to explore good lecture practices in Chapter 8) to understand how active learning approaches are beneficial. Fig. 4.1 illustrates the model, which suggests that activities that foster connections between selected new information and existing knowledge build long-term memory. The kinds of activities that can foster these connections involve interpretation, or the process of fitting new information with what is known, and elaboration, which

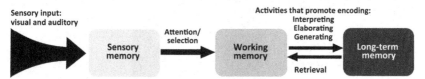

FIGURE 4.1 A model of memory formation can help explain the importance of active learning. Some sensory input is selected for processing in working memory, which has low capacity and is short term. Connecting information in working memory to existing knowledge leads to encoding in long-term memory. *The model shown is based on the Atkinson–Shiffrin model, supplemented with information from deWinstanley and Bjork* (Atkinson and Shiffrin, 1968; deWinstanley and Bjork, 2002). *Although the model can be useful for making teaching choices, it's important to note that it's highly simplified. For example, short-term working memory and long-term memory function as a continuum rather than as two distinct entities.*

extends interpretation to incorporate new information into a broader, coherent narrative and to consider its implications. The specific activities that promote interpretation and elaboration can vary enormously in structure, from pauses to allow students to put together notes and identify questions and areas of confusion, to prompts that encourage students to predict an outcome or construct a model from seemingly disconnected pieces of information.

IS THERE EVIDENCE THAT IT WORKS?

The evidence that active learning is effective is robust and stretches back more than thirty years (see e.g., Bonwell and Eison, 1991). Here, we will focus on two reports that review and analyze multiple active learning studies.

Freeman and colleagues conducted a meta-analysis of 225 studies comparing "constructivist versus exposition-centered course designs" in STEM disciplines (Freeman et al., 2014). They included studies that examined the design of class sessions (as opposed to out-of-class work or laboratories) with at least some active learning versus traditional lecturing, comparing failure rates and student scores on examinations, concept inventories, or other assessments. They found that students in traditional lectures were 1.5 times more likely to fail than students in courses with active learning (odds ratio of 1.95, $Z = 10.4$, $P \ll 0.001$; Fig. 4.2). Further, they found that on average, student performance on exams, concept inventories, or other assessments increased by about half a standard deviation when some active learning was included in course design (weighted standardized mean difference of 0.47,

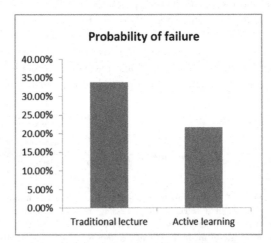

FIGURE 4.2 **Active learning reduces the probability of failure when compared to traditional lecture.** A meta-analysis of 225 studies comparing "constructivist versus exposition-centered course designs" in STEM disciplines found that the failure rate decreased from about 34% to about 22%. Figure generated based on data reported in Freeman et al. (2014).

$Z = 9.781$, $P \ll 0.001$). These results were consistent across disciplines: they observed no significant difference in the effects of active learning in biology, chemistry, computer science, engineering, geology, math, physics, and psychology courses. They performed two analyses examining the possibility that the results were due to a publication bias (i.e., a bias toward publishing studies with larger effects), finding that there would have to be a large number of unpublished studies that observed no difference between active learning and lecturing to negate their findings: 114 reporting no difference on exam or concept inventory performance and 438 reporting no difference in failure rate. The authors conclude that the evidence for the benefits of active learning are very strong, stating that, "If the experiments analyzed here had been conducted as randomized controlled trials of medical interventions, they may have been stopped for benefit—meaning that enrolling patients in the control condition might be discontinued because the treatment being tested was clearly more beneficial."

These results support other, earlier reviews (e.g., Hake, 1998; Prince, 2004; Springer et al., 1999). In one such review, Ruiz-Primo and colleagues examined published studies examining the effects of active learning approaches in undergraduate biology, chemistry, engineering and physics courses (Ruiz-Primo et al., 2011). They identified 166 studies that reported an effect size when comparing the effects of an innovation (i.e., active learning approaches) to traditional instruction that did not include the innovation. Overall, they found that inclusion of the active learning approaches improved student outcomes (mean effect size = 0.47), although there are important caveats to consider. First, the authors coded the active learning activities as conceptually oriented tasks, collaborative learning activities, technology-enabled activities, inquiry-based projects, or some combination of those four categories, and important differences existed within the categories (e.g., technology-assisted inquiry-based projects on average did not produce positive effects). Second, more than 80% of the studies included were quasi-experimental rather than experimental, and the positive benefits were lower for the experimental studies in which students were randomly assigned to a treatment group (mean effect size = 0.26). Finally, many of the studies did not control for preexisting knowledge and abilities in the treatment groups. Notwithstanding these caveats, the review provides support for the inclusion of active learning approaches in instruction.

WHY IS IT IMPORTANT? MAKING YOUR CLASS MORE INCLUSIVE

In addition to the evidence that active learning approaches promote learning for all students, there is provocative and compelling evidence that active learning approaches are an effective tool in making classrooms more inclusive. Haak and colleagues examined the effects of active learning for students in the University of Washington's Educational Opportunity Program

(EOP) who were enrolled in an introductory biology course (Haak et al., 2011). Students in the EOP are educationally or economically disadvantaged, are typically the first in their families to attend college, and include most underrepresented minority students at the University of Washington. Previous work had demonstrated that the researchers could predict student grades in the introductory biology course based on their college grade point average (GPA) and Scholastic Aptitude Test (SAT) verbal score; students in the EOP had a mean failure rate of $\sim 22\%$ compared to a mean failure rate of $\sim 10\%$ for students not in the EOP. When multiple highly structured approaches to promote active learning were incorporated into the introductory biology course, all students in the course benefited, but students in the EOP demonstrated a disproportionate benefit, reducing the achievement gap to almost half of the starting level.

Eddy and Hogan also found that active learning resulted in disproportionate benefits for black and first-generation students in an introductory science class. They compared student performance in an introductory biology class taught either through traditional lecture, which they defined as a course in which students talked for <15% of class time and did fewer than one out-of-class assignments per week, or through a course with moderate structure, characterized by students talking 15−40% of class time and completing one out-of-class assignment per week (Eddy and Hogan, 2014). Importantly, they looked at the impact of increased structure on different groups of students in the course: white, Native American, Asian, black, Latinx, mixed race, and international; male and female; and first- and continuing-generation college students. All students demonstrated stronger performance in the course with increased structure, but the greatest gains were seen for black and first-generation students. They also investigated the causes for the improved performance, finding that adding active learning exercises to the course (both in and out of class) increased student-reported study time, and increased in-class participation for black students.

Ballen and colleagues also investigated the impact of incorporating in- and out-of-class active learning approaches in an introductory biology class (Ballen et al., 2017). This group found that increasing active learning eliminated a performance gap between underrepresented minority (URM) and majority students. Importantly, the group found that an increase in students' science self-efficacy mediated the increase in performance for URM students.

Lorenzo, Crouch, and Mazur also investigated the impact of active learning approaches on the difference in male and female performance in introductory physics classes (Lorenzo et al., 2006). They found that inclusion of active engagement techniques benefited all students, but had the greatest impact on female students' performance. In fact, when they included a "high dose" of active learning approaches, the gender gap was eliminated. This result supports earlier work suggesting that women particularly benefit from active learning approaches (Laws et al., 1999; Schneider, 2001).

Given the pressing need to make U.S. college science classrooms more inviting and productive spaces for students from all backgrounds, these results provide compelling reasons to incorporate active learning approaches into course design.

WHAT ARE TECHNIQUES TO USE?

Brief, Easy Supplements to Lecture

The pause procedure. Pause for 2 minutes every 12−18 minutes, encouraging students to discuss and rework notes in pairs. This approach encourages students to consider their understanding of the lecture material, including its organization. It also provides an opportunity for questioning and clarification and has been shown to significantly increase learning when compared to lectures without the pauses. (Bonwell and Eison, 1991; Rowe, 1980; Ruhl et al., 1987)

Retrieval practice. Pause for 2 or 3 minutes every 15 minutes, having students write everything they can remember from preceding class segment. Encourage questions. This approach prompts students to retrieve information from memory, which improves long-term memory, ability to learn subsequent material, and ability to translate information to new domains. (Roediger and Butler, 2011; Brame and Biel, 2015; for more ideas on how to use this approach, see Chapter 7).

Think−pair−share. Ask students a question that requires higher-order thinking (e.g., application, analysis, or evaluation levels within Bloom's taxonomy; see Fig. 4.3). Ask students to think or write about an answer for

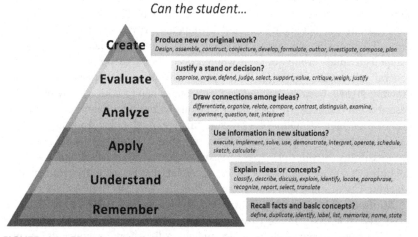

Can the student...

Create — Produce new or original work?
Design, assemble, construct, conjecture, develop, formulate, author, investigate, compose, plan

Evaluate — Justify a stand or decision?
appraise, argue, defend, judge, select, support, value, critique, weigh, justify

Analyze — Draw connections among ideas?
differentiate, organize, relate, compare, contrast, distinguish, examine, experiment, question, test, interpret

Apply — Use information in new situations?
execute, implement, solve, use, demonstrate, interpret, operate, schedule, sketch, calculate

Understand — Explain ideas or concepts?
classify, describe, discuss, explain, identify, locate, paraphrase, recognize, report, select, translate

Remember — Recall facts and basic concepts?
define, duplicate, identify, label, list, memorize, name, state

FIGURE 4.3 Bloom's revised taxonomy describes six levels of cognitive processes. Figure from Vanderbilt University Center for Teaching; https://cft.vanderbilt.edu/guides-sub-pages/blooms-taxonomy/.

1 minute, then turn to a peer to discuss their responses for 2 minutes. Have selected groups share their responses with the whole class and follow up with instructor explanation. By asking students to explain their answer to a neighbor and critically to consider their neighbor's responses, this approach helps students articulate newly formed mental connections.

Peer instruction. This modification of the think−pair−share involves personal response devices (e.g., clickers). Pose a conceptually based multiple-choice question. Ask students to think about their answer and vote on a response before turning to a neighbor to discuss. Encourage students to change their answers after discussion, if appropriate, and share class results by revealing a graph of student responses. Use the graph as a stimulus for class discussion. This approach is particularly well-adapted for large classes and can be facilitated with a variety of tools (e.g., Poll Everywhere, TopHat, TurningPoint). *LSE* provides an evidence-based teaching guide, found at lse.ascb.org/evidence-based-teaching-guides/peer-instruction/, that can help guide instructor choices when using this approach (Fagen et al., 2002; Crouch and Mazur, 2001).

Minute papers. Ask students a question that requires them to reflect on their learning or to engage in critical thinking. Have them write for 1 minute. Ask students to share responses to stimulate discussion or collect all responses to inform future class sessions. Like the think−pair−share approach, this approach encourages students to articulate and examine newly formed connections (Angelo and Cross, 1993; Handelsman et al., 2007).

Activities to Replace Some Lecture

Concept map. Concept maps are visual representations of the relationships between concepts. Concepts are placed in nodes (often, circles), and the relationships between indicated by labeled arrows connecting the concepts. To have students create a concept map, identify the key concepts to be mapped in small groups or as a whole class. Ask students to determine the general relationship between the concepts and arrange them two at a time, drawing arrows between related concepts and labeling with a short phrase to describe the relationship. By asking students to build an external representation of their mental model of a process, this approach helps students examine and strengthen the organization within the model. Further, it can emphasize the possibility of multiple "right" answers. More information and a tool to do online concept mapping can be found at the Institute for Human and Machine Cognition (Novak and Cañas, 2008).

Mini-maps. Mini-maps are like concept maps, but students are given a relatively short list of terms (usually 10 or fewer) to incorporate into their map. To use this approach, provide students a list of major concepts or specific terms and ask them to work in groups of two or three to arrange the terms in a logical structure, showing relationships with arrows and words.

Ask groups to volunteer to share their mini-maps and clarify any confusing points. Mini-maps have many of the same strengths as concept maps but can be completed more quickly and thus can serve as part of a larger class session with other learning activities (Handelsman et al., 2007).

Categorizing grids. Present students with a grid made up of several important categories and a list of scrambled terms, images, equations, or other items. Ask students quickly to sort the terms into the correct categories in the grid. Ask volunteers to share their grids and answer questions that arise. This approach allows students to express and thus interrogate the distinctions they see within a field of related items. It can be particularly effective at helping instructors identify misconceptions (Angelo and Cross, 1993).

Student-generated test questions. Provide students with a copy of your learning goals for a particular unit and a figure summarizing Bloom's taxonomy with representative verbs associated with each category (such as the one shown in Fig. 4.3). Challenge groups of students to create test questions corresponding to your learning goals and different levels of the taxonomy. Consider having each group share its favorite test question with the whole class or consider distributing all student-generated questions to the class as a study guide. This approach helps students consider what they know as well as implications of the instructor's stated learning goals (Angelo and Cross, 1993).

Decision-making activities. Ask students to imagine that they are policymakers who must make and justify tough decisions. Provide a short description of a thorny problem, ask them to work in groups to arrive at a decision, and then have groups share their decisions and explain their reasoning. This highly engaging technique helps students critically consider a challenging problem and encourages them to be creative in considering solutions. The "real-world" nature of the problems can provide incentive for students to dig deeply into the problems. An example is shown in Box 4.2 (Handelsman et al., 2007).

Case-based learning. Much like decision-making activities, case-based learning presents students with situations from the larger world that require students to apply their knowledge to reach a conclusion about an open-ended situation. Provide students with a case, asking them to decide what they know that is relevant to the case, what other information they may need, and

---

BOX 4.2 Example Decision-Making Activity

You are the head of a major blood bank, and there is a worldwide blood shortage. You are offered a shipment of blood that might be contaminated with a new retrovirus that has not been well-studied. Will you allow the blood to be used? Why? What would you like to know before you make your decision?

Handelsman et al. (2007).

what impact their decisions may have, considering the broader implications of their decisions. Give small groups (3—5) of students time to consider responses, circulating to ask questions and provide help as needed. Provide opportunities for groups to share responses; the greatest value from case-based learning comes from the complexity and variety of answers that may be generated. More information and collections of cases are available at the National Center for Case Study Teaching in Science.

Demonstrations. Ask students to predict the result of a demonstration, briefly discussing with a neighbor. After demonstration, ask them to discuss the observed result and how it may have differed from their prediction; follow up with instructor explanation. This approach asks students to test their understanding of a system by predicting an outcome. If their prediction is incorrect, it helps them see the misconception and thus prompts them to restructure their mental model.

Strip sequence. Give students the steps in a process on strips of paper that are jumbled; ask them to work together to reconstruct the proper sequence. This approach can strengthen students' logical thinking processes and test their mental model of a process (Handelsman et al., 2007).

Other Approaches

There are other active learning pedagogies, many of which are highly structured and have dedicated websites and strong communities. These include team-based learning (TBL), process-oriented guided inquiry learning (POGIL), peer-led team learning, and problem-based learning (PBL). Further, the flipped classroom model is based on the idea that class time will be spent with students engaged in active learning; this approach is explored in Chapter 9.

HOW SHOULD YOU GET STARTED?

Start small, start early, and start with activities that pose low risk for both instructors and students. The pause procedure, retrieval practice, minute papers, and the think—pair—share technique provide easy entry points to incorporating active learning approaches, requiring the instructor to change very little while providing students an opportunity to organize and clarify their thinking. As you begin to incorporate these practices, it's a good idea to explain to your students why you're doing so; talking to your students about their learning not only helps build a supportive classroom environment, but can also help them develop their metacognitive skills (and thus their ability to become independent learners).

As you consider other active learning techniques to use, use the "backwards design" approach: begin by identifying your learning goals, think about how you would identify whether students had reached them (that is, how you might structure assessment), and then choose an active learning approach that helps your students achieve those goals. Students typically have positive responses to active learning activities that are meaningful, appropriately challenging, and clearly tied to learning goals and assessments (see e.g., Lumpkin et al., 2015). These points are worth emphasizing: active learning exercises that are too easy or are unrelated to learning goals and assessments will seem like busywork to students and will not be valued. Choosing your questions and problems to fit the course and its assessments is key for success.

Finally, use your local community: consult colleagues within your department and/or your institution's Center for Teaching and Learning for help and feedback as you design and implement active learning approaches.

CONCLUSION

In this section of the book, we are exploring four adaptable teaching and learning approaches that can be combined and adapted to form the basis of the 21st century science class. Active learning is one of these cornerstones, serving as a tool to help students make the connections they need to develop a rich understanding of our course material. Many active learning exercises involve cooperative learning in small groups, and for good reason: Johnson et al. (2014) have shown that cooperative learning is one of the most robust means to improve student learning in higher education, and several studies have illustrated its value in active learning (e.g., Linton et al., 2014; Smith et al., 2011). The next chapter turns specifically to group work, providing guidance on how to maximize the benefits and avoid the pitfalls of this key teaching and learning approach.

REFERENCES

Angelo, T.A., Cross, K.P., 1993. Classroom Assessment Techniques: A Handbook for College Teachers. Jossey-Bass, San Francisco, CA.

Atkinson, R.C., Shiffrin, R.M., 1968. Human memory: a proposed system and its control processes. In: Spence, K.W., Spence, J.T. (Eds.), The Psychology of Learning and Motivation, vol. 2. Academic Press, New York, NY, pp. 89−195.

Ballen, C.J., Wieman, C., Salehi, S., Searle, J.B., Zamudio, K.R., 2017. Enhancing diversity in undergraduate science: self-efficacy drives performance gains with active learning. CBE Life Sci. Educ. 16 (4), ar56.

Bonwell, C.C., Eison, J.A., 1991. Active Learning: Creating Excitement in the Classroom. 1991 ASHE-ERIC Higher Education Report No. 1. The George Washington University. School of Education and Human Development, Washington, DC.

Brame, C.J., Biel, R., 2015. Test-enhanced learning: the potential for testing to promote greater learning in undergraduate science courses. CBE Life Sci. Educ. 14, 1−12.

Bransford, J.D., Brown, A.L., Cocking, R.R. (Eds.), 1999. How People Learn: Brain, Mind, Experience, and School. National Academy Press, Washington, DC.

Carr, R., Palmer, S., Hagel, P., 2015. Active learning: the importance of developing a comprehensive measure. Act. Learn. High. Educ. 16, 173–186.

Crouch, C.H., Mazur, E., 2001. Peer instruction: ten years of experience and results. Am. J. Phys. 69, 970–977.

deWinstanley, P.A., Bjork, R.A., 2002. Successful lecturing: presenting information in ways that engage effective processing. New Dir. Teach. Learn. 89, 19–31.

Eddy, S.L., Hogan, K.A., 2014. Getting under the hood: how and for whom does increasing course structure work? CSE Life Sci. Educ. 13, 453–468.

Fagen, A.P., Crouch, C.H., Mazur, E., 2002. Peer instruction: results from a range of classrooms. Phys. Teach. 40, 206–209.

Freeman, S., Eddy, S.L., McDonough, M., Smith, M.K., Okoroafor, N., Jordt, H., et al., 2014. Active learning increases student performance in science, engineering, and mathematics. Proc. Nat. Acad. Sci. U.S.A. 111, 8410–8415.

Haak, D.C., HilleRisLambers, J., Pitre, E., Freeman, S., 2011. Increased structure and active learning reduce the achievement gap in introductory biology. Science 332, 1213–1216.

Hake, R., 1998. Interactive-engagement versus traditional methods: a six-thousand-student survey of mechanics test data for introductory physics courses. Am. J. Phys. 66, 64–74.

Handelsman, J., Miller, S., Pfund, C., 2007. Scientific Teaching. W.H. Freeman, New York, NY.

Johnson, D.W., Johnson, R.T., Smith, K.A., 2014. Cooperative learning: Improving university instruction by basing practice on validated theory. J. Excell. Coll. Teach. 25, 85–118.

Laws, P., Rosborough, P., Poodry, F., 1999. Women's responses to an activity-based introductory physics program. Am. J. Phys. 67, S32–S37.

Linton, D.L., Farmer, J.K., Peterson, E., 2014. Is peer interaction necessary for optimal active learning? CBE Life Sci. Educ. 13, 243–252.

Lorenzo, M., Crouch, C.H., Mazur, E., 2006. Reducing the gender gap in the physics classroom. Am. J. Phys. 74, 118–122.

Lumpkin, A., Achen, R., Dodd, R., 2015. Student perceptions of active learning. Coll. Stud. J. 49, 121–133.

Novak, J.D., Cañas, A.J., 2008. The theory underlying concept maps and how to construct and use them. Technical Report IHMC CmapTools 2006-01 Rev 2008-01, Institute for Human and Machine Cognition. Retrieved from: http://cmap.ihmc.us/docs/theory-of-concept-maps.

Prince, M., 2004. Does active learning work? A review of the research. J. Eng. Educ. 93, 223–231.

Roediger, H.L., Butler, A.C., 2011. The critical role of retrieval practice in long-term retention. Trends Cogn. Sci. 15, 20–27.

Rowe, M.B., 1980. Pausing principles and their effects on reasoning in science. New Dir. Commun. Coll. 31, 27–34.

Ruhl, K., Hughes, C.A., Schloss, P.J., 1987. Using the pause procedure to enhance lecture recall. Teach. Educ. Spec. Educ. 10, 14–18.

Ruiz-Primo, M.A., Briggs, D., Iverson, H., Talbot, R., Shepard, L.A., 2011. Impact of undergraduate science course innovations on learning. Science 331, 1269–1270.

Schneider, M., 2001. Encouragement of women physics majors at Grinnell College: a case study. Phys. Teach. 39, 280–282.

Smith, M.K., Wood, W.B., Krauter, K., Knight, J.K., 2011. Combining peer discussion with instructor explanation increases student learning from in-class concept questions. CSE Life Sci. Educ. 10, 55–63.

Springer, L., Stanne, M.E., Donovan, S.S., 1999. Effects of small-group learning on undergraduates in science, mathematics, engineering, and technology. Rev. Educ. Res. 69, 21–51.

Chapter 5

Group Work: Using Cooperative Learning Groups Effectively

Group work is the Dr. Jekyll/Mr. Hyde of the college classroom. When it goes well, it can lead to very positive experiences that let students learn from and teach their colleagues as they tackle problems harder than they can manage alone. When it goes poorly, it can produce little learning and lots of frustration and resentment. And most of us recognize that there are multiple ways group work can go poorly: students not contributing, students dominating, students arguing within the group, students not engaging with the problem.

The challenge is to construct group work to maximize the probability that students will engage and learn. This chapter provides guidance on how to structure group work so that it is a positive and productive experience. It describes the differences between informal and formal group work, evidence that group work can improve students' learning, and instructional choices that can maximize the benefits of these two types of group work.

WHAT CAN IT LOOK LIKE?

An easy and highly effective way to use group work is to pose questions or problems in class to small, temporary, ad hoc groups of two to four students. The students work together for brief periods, from 2 to 3 minutes to a single class period, to answer questions or respond to prompts posed by the instructor. These informal groups are an integral part of many active learning techniques. The barriers for engaging in this type of group work are low for both instructors and students, and benefits can be large and varied, from addressing common misconceptions to building community in the classroom. Fig. 5.1 gives a few ways to structure informal group work.

If we want our students to tackle a larger project that takes more time, it can be useful to form longer- lasting groups that have a formal commitment to completing a project together. Because the group tasks are larger and the stakes are higher, it is important to be particularly careful in setting up this type of group work. Many well-known pedagogies frequently used in science classes, such as team-based learning, problem-based learning,

Science Teaching Essentials. DOI: https://doi.org/10.1016/B978-0-12-814702-3.00005-6

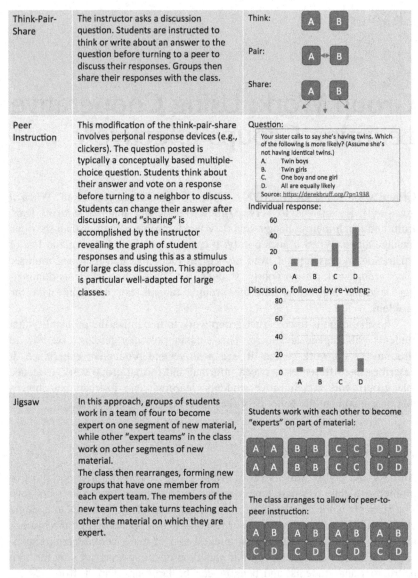

Think-Pair-Share	The instructor asks a discussion question. Students are instructed to think or write about an answer to the question before turning to a peer to discuss their responses. Groups then share their responses with the class.	Think: Pair: Share:
Peer Instruction	This modification of the think-pair-share involves personal response devices (e.g., clickers). The question posted is typically a conceptually based multiple-choice question. Students think about their answer and vote on a response before turning to a neighbor to discuss. Students can change their answer after discussion, and "sharing" is accomplished by the instructor revealing the graph of student responses and using this as a stimulus for large class discussion. This approach is particular well-adapted for large classes.	Question: Your sister calls to say she's having twins. Which of the following is more likely? (Assume she's not having identical twins.) A. Twin boys B. Twin girls C. One boy and one girl D. All are equally likely Source: https://derekbruff.org/?p=1938 Individual response: Discussion, followed by re-voting:
Jigsaw	In this approach, groups of students work in a team of four to become expert on one segment of new material, while other "expert teams" in the class work on other segments of new material. The class then rearranges, forming new groups that have one member from each expert team. The members of the new team then take turns teaching each other the material on which they are expert.	Students work with each other to become "experts" on part of material: The class arranges to allow for peer-to-peer instruction:

FIGURE 5.1   Examples of informal cooperative learning activities.

and process-oriented guided inquiry learning, require this type of more formalized group work.

Since using group work requires careful thought about appropriate tasks and group structures, it is important to examine some of the evidence that it is worth the effort—that is, that group work can enhance learning when compared to individualistic learning. The next section of the chapter summarizes some of the research on the effectiveness of group work, and the final sections then provide specific recommendations for both informal and formal group work.

IS THERE EVIDENCE THAT IT PROMOTES LEARNING?

When students work together to solve instructor-defined problems and to reach instructor-defined goals, they are participating in a process formally termed cooperative learning. Cooperative learning is defined as the instructional use of small groups to promote students working together to maximize their own and each other's learning (Johnson and Johnson, 2009). It is characterized by positive interdependence, where students perceive that better performance by individuals produces better performance by the entire group (Johnson et al., 2014).

Johnson et al. (2007) performed a meta-analysis of 168 studies comparing cooperative learning to competitive learning and individualistic learning in college students, where they defined competitive learning as students working against each other to achieve an academic goal, and individualistic learning as students working by themselves to accomplish learning goals unrelated to those of other students. They found that cooperative learning produced greater academic achievement than both competitive learning and individualistic learning across the studies, exhibiting a mean weighted effect size of 0.54 when comparing cooperation and competition and 0.51 when comparing cooperation and individualistic learning. In essence, these results indicate that cooperative learning increases student academic performance by approximately one-half of a standard deviation when compared to noncooperative learning models, an effect that is considered moderate. Importantly, each study included in the meta-analysis defined the academic achievement measures, and these ranged from lower-level cognitive tasks (e.g., knowledge acquisition and retention) to lower-level cognitive activity (e.g., creative problem-solving), and from verbal tasks to mathematical tasks to procedural tasks. The meta-analysis also showed substantial effects on other metrics, including self-esteem and positive attitudes about learning. Kuh et al. (2007) also conclude that cooperative group learning promotes student engagement and academic performance.

Springer et al. (1999) found similar results in their meta-analysis of 39 studies of small-group learning in postsecondary STEM courses from 1980 to 1999. They found that students who participated in various types of small-group learning, ranging from extended formal interactions to brief informal interactions, had greater academic achievement, exhibited more favorable attitudes toward learning, and had increased persistence through STEM courses than students who did not participate in STEM small-group learning. They also examined the effects of small-group learning based on gender and racial or ethnic composition of the group, finding little evidence that gender composition altered outcomes but observing greater benefits for groups composed primarily or exclusively of African-American or Latinx students.

It is important to note that many of the studies included in these meta-analyses explicitly examined the impact of cooperative learning—not just throwing students into groups and telling them to work together; but, instead, choosing group structures, group activities and tasks, and reward systems that promoted students' efforts to help each other learn. The following recommendations can help you structure your group work to achieve this goal.

WHAT INSTRUCTIONAL CHOICES CAN HELP INFORMAL GROUP WORK BE EFFECTIVE?

Using informal group work in class can be fairly simple; you can ask a question, give students 30 seconds to think or write about it, and then have them discuss their response with a neighbor before reporting out to the larger group. There are choices to make, however. What types of questions are effective? Should work be graded, and, if so, for completion or correctness? Can the instructor impact whether all students express their ideas in the discussions with peers, and does it matter? After students have the opportunity to discuss answers with their colleagues, should the instructor "cold-call" groups to report out or ask for volunteers? Is it helpful to use personal response devices to solicit responses from all groups? Many of these questions have been addressed in the context of peer instruction, a pedagogy based on the traditional think—pair—share approach that is diagrammed in Fig. 5.1. Recommendations by this research are summarized here, and CBE—Life Sciences Education provides an evidence-based teaching guide with summaries of and links to many relevant articles (lse.ascb.org/evidence-based-teaching-guides/peer-instruction/).

1. *Questions should be tied to your learning goals and should help students with tests or other assessments.* One of the most important elements for ensuring that students value the work they do in your class is alignment between your learning goals and the ways students earn grades. If students perceive that the questions they are tackling with their colleagues in class are important and help them perform better on later tests, then they will see this work as an integral part of the class and will engage. If, on the other hand, the questions seem to be unrelated to the core learning goals of the class as reflected by the way students are graded, then they will see the activities as superfluous and will often choose not to engage. (If you are tempted to critique students for this attitude, consider your own tendencies in faculty development sessions; I predict that if you think an activity will help you in your classes, you are more likely to engage in a meaningful way than if you are learning just for fun.)

2. *Questions or prompts should be challenging but accessible.* Tasks that are challenging but doable enhance learning, an observation that led Robert Bjork to coin the phrase "desirable difficulty" (Bjork and Bjork, 2011). This observation appears to be true for more than a single reason. First, challenging tasks are more interesting and likely to engage students. Second, challenging tasks promote the sort of processing that is needed to link new information to existing knowledge and thus to help create the network of mental associations that we want our students to build. In addition, there is something of a combination of these effects when we are considering group work: students are more likely to work with each other to explain their reasoning and confusion with a challenging task, thereby engaging in the sort of interpretation and elaboration they need to do to encode memories. Much of the literature on desirable difficulties has investigated the role of spacing practice and alternating topics to create difficulties, but the task itself also has the potential to generate this effect (McDaniel and Butler, 2011; Smith et al., 2009). It is also important that the task be accessible: tasks that are too hard can make students feel helpless and unwilling to engage. A good source for questions may be your own or colleagues' previous tests—choose a couple of challenging questions and pose them to your class, encouraging students to work together to come to an answer.

3. *Question format can vary.* There is no right or wrong type of question to stimulate small-group work. Challenging multiple choice questions— either with a single or multiple correct answers—can be an effective way to engage large classes with limited seating options (Crouch and Mazur, 2001). Questions that ask students to draw a concept map, a model, or a structure can also engage students, and can be adapted for large classes by using small, portable whiteboards. Questions that ask students to generate a verbal or written explanation of a phenomenon can also work well. As long as the questions are aligned with your learning goals and assessments and are challenging but accessible, format can vary.

4. *Low-stakes grading increases student engagement and participation.* It is somewhat disappointing, but giving students credit for participating in small-group work increases the probability that all students will participate and results in more robust exchanges of reasoning when students are discussing possible answers (Freeman et al., 2007; James, 2006; James et al., 2008). It appears important that grading be low-stakes; however, higher stakes grading that emphasizes correct answers appears to promote discussions where one student dominates the group.

5. *Instructor guidance can increase student contributions to peer discussions.* Instructor framing of small-group work is critically important. Many of our students are comfortable working alone and may be hesitant to talk about tasks with their colleagues, perhaps because they are shy, do not want to take the chance of looking ignorant, or do not see benefits.

By explaining the benefits of working with colleagues and how the processing students are doing builds memory and understanding—and how this will help students on subsequent tests or other assessments—the instructor can help students value this work. (This can also help position the instructor as an expert teacher, which can enhance students' trust and efforts in the class.) In addition, it is important to encourage students to explain their reasoning as they go about the group work task (Knight et al., 2013). Articulating their reasoning to colleagues helps students make the kinds of connections they need to build memory and reveals misconceptions and misunderstandings they may have. Finally, it can be helpful for instructors to be specific about students taking turns talking. Although low-stakes grading increases student engagement and reduces domination by a single student, groups will often have a student who tends to talk more—and this student gets the greatest benefit from group work. By specifying that it is time for a switch in speaker, the instructor can help ensure that all students get the benefit of articulating their thinking processes (Kagan, 2014).

6. *Cold-calling students after peer discussion has both benefits and drawbacks.* Knight and colleagues have demonstrated that group random call after peer discussion increases students' focus on reasoning during discussion (Knight et al., 2016). Further, Eddy and colleagues found that relying on student volunteers for answers to questions in introductory biology classes leads to an underrepresentation of women's voices (Eddy et al., 2014). However, random call may also promote student anxiety (Cooper et al., 2018). If you do choose randomly to call on groups, it can be helpful to emphasize that you are looking for a group answer, that the reasoning behind the answer is as important as the answer itself, and that revealing misunderstandings has benefits for learning.

7. *Students like the use of personal response devices, but their use does not appear to impact learning.* Students often find answering questions on personal response devices (such as with iClicker or TopHat) engaging, and it provides an easy way to introduce low-stakes grading (e.g., Stowell and Nelson, 2007). There is no evidence, however, that the use of the technology enhances students' learning from group work.

WHAT INSTRUCTIONAL CHOICES CAN HELP FORMAL GROUP WORK BE EFFECTIVE?

Informal group work is a great tool to have in your toolkit, and it can be used in almost any class session. Sometimes, however, we want our students to work in groups to tackle bigger projects over multiple days or weeks. You have multiple choices when you are setting up a group project like this for your students. How big should the groups be? Should you form them, and, if so, using what criteria? Can you do anything to help groups work together well? What should the task be? How should grades be assigned? The bigger and longer-term the project is, the more important the answers to these questions are. Many of these questions have been addressed in research studies, allowing you to make evidence-based choices for your classroom.

Recommendations that are supported by current research are summarized here, and *CBE—Life Sciences Education* provides an evidence-based teaching guide with summaries of and links to many relevant articles (lse.ascb. org/evidence-based-teaching-guides/group-work/).

1. *Smaller teams tend to work better.* Most studies that look at group size find that groups of 3—5 produce greater cooperativity and less social loafing, resulting in better group performance and higher satisfaction (Lou et al., 2001; Treen et al., 2016; Heller and Hollabaugh, 1992).

2. *There is conflicting evidence about whether it is better for instructors to form groups or for students to self-select groups.* Feichtner and Davis (1984) found that self-selected teams were more likely to be given as examples of students' worst group experiences, but other studies have found that students prefer self-selection over random assignment in advanced courses and that they experience higher levels of communication, commitment, and satisfaction in self-selected teams (Chapman et al., 2006; Myers, 2012). Further, Eddy et al. (2015) found that female students perceived greater value in peer-to-peer discussion when they had a friend in the group. On balance, therefore, it may be better to have students self-select groups in more advanced courses where they know other students, but for instructors to assign groups and to include team-building activities in courses where students do not know each other (see more later).

3. *It is important to provide an opportunity for students to discuss their expectations for group work, to set group norms, and to plan how they are going to work together* (Johnson et al., 2007). Asking students to discuss their concerns about group work within their teams can help them use their prior experience to identify ways to avoid pitfalls. It can be particularly helpful to encourage your students to talk about common problems, such as students dominating or not contributing, and to consider how to head off those problems. Lerner suggests providing students with names and descriptions that correspond to common problems, and to have student teams plan a response. The humor associated with planning for Nola No-Can-Meet, Do-It-All Dottie, Seldom-Seen Steve, and Always-Right Artie can help take the edge off these discussions while still providing an opportunity for the group to come together as a team (Lerner, 1995). It can also be helpful to have students make a plan for the project—identifying tasks, naming responsible team members, predicting the length of time it will take—that can serve as a contract with each other and can also give you checkpoints to monitor.

4. *The group task should be complex or loosely defined, enhancing the value of group collaboration.* Many of our students are highly motivated, well prepared, and comfortable with working alone. If we assign a group task that is too simple, we provide little incentive for students to work together to solve the problem. Instead, it becomes more likely that a single student in the group will contribute most to the project. The contributing student learns no more than he or she would have working alone, and the others learn less. Tasks that are dense and complex, requiring input from multiple group members, promote contribution from the team (Scager et al., 2016; Kirschner et al., 2011). There are multiple resources for complex problems that instructors can use or adapt for their classes:

   • The Problem-Based Learning website from University of Delaware (www1.udel.edu/inst/) provides a clearinghouse of ill-structured problems and other resources.

   • The National Center for Case Study Teaching in Science (sciencecases.lib.buffalo.edu/cs/) provides >700 cases, ranging from relatively simple to more advanced.

   • Process Oriented Guided Inquiry Learning (POGIL) is a pedagogy that has been widely used in chemistry. The POGIL website (pogil. org) provides a few examples of activities and links to several textbooks that provide more.

5. *Tasks that tap into students' intrinsic motivation increase the probability of student investment and collaboration* (Scager et al., 2016; Allchin, 2013; Schmidt et al., 2011). Student motivation is absolutely essential to any learning activity, and may be more important for group projects. Student motivation can be tied to student interest—that is, whether students are curious or care about the problem they are presented with. Tasks that are contextualized, such as the problems and cases provided at the websites above, often help stoke this type of student interest, helping students see the value in the content they are learning and problems they are addressing. Student motivation can also be tied to the nature of the final product. Products that have value beyond the class, such as articles that are written for publication or posters that are presented to a broader audience, provide another means to promote student motivation.

6. *Grading structures that assess individual contributions as well as the group product help promote equitable contributions and team cooperativity* (Sears and Pai, 2012; Brewer and Klein, 2006; Serrano and Pons, 2007). This approach is most easily illustrated with testing examples. In one example of a grading structure that promoted high levels of positive interdependence, helping behavior, communication, and deep learning approaches, students received 30% of their grade from their individual performance on a test and 70% of their grade from the mean score obtained by their group (Serrano and Pons, 2007). For projects that do not lead to testing, having students identify their individual contributions while also giving significant weight to the collective final project can be

powerful. To avoid "Frankenprojects" that look like individual parts stitched together, it is important to emphasize in your instructions and grading that the final project should be cohesive.

7. *Peer evaluation can be a powerful tool for promoting accountability.* Giving students the opportunity to evaluate the contributions of team members can lead to more positive attitudes about group work (Chapman and Van Auken, 2001). It can also help the instructor capture information about team members' contributions that may not be obvious from viewing the final project. In addition to asking for comments about what team members are doing well and where they need to improve, instructors can also give students a set number of points to distribute among team members. If, for example, a student has 25 points to distribute to three teammates (not including herself), a distribution of 8, 8, and 9 indicates a fairly equitable contribution by the three coworkers. If one student contributed significantly more, however, one might see a distribution of 7, 7, and 11. By averaging the number given to any given student, the instructor can typically get a sense of the contributions that team member made, supported by the students' written comments. This approach gives students the opportunity to be concrete with their feedback, differentiating among teammates who contribute and those who do not. If students know that they will be evaluated in this way, they are more likely to be conscientious about being good group members. Wenzel (2007) provides several other tools for peer- and self-assessment for class- and lab-based projects.

CONCLUSION

Making thoughtful, informed choices about the group work that you use in your classes can help you capitalize on the potential of this approach. From using transient small groups to help students confront their misconceptions during lecture to having teams work together on larger design or research projects, instructors can use cooperative groups to help their students achieve the kind of learning they want. The recommendations offered here can help you determine which type of group work can best achieve your goals as well as help you make instructional choices that foster productive interactions in the student teams.

In this section of the book, we have considered two important, adaptable teaching approaches: active learning and group work. One benefit that both of these approaches have is that they give students tools to articulate and monitor their understanding, which is a key part of metacognition. The next chapter specifically focuses on approaches that we can use to help our students become more metacognitive. These approaches align with and can be integrated into many teaching approaches, giving students multiple tools to enhance their learning.

REFERENCES

Allchin, D., 2013. Problem- and case-based learning in science: an introduction to distinctions, values, and outcomes. CBE Life Sci. Educ. 12, 364–372.

Bjork, E.L., Bjork, R.A., 2011. Making things hard on yourself, but in a good way: creating desirable difficulties to enhance learning. In: Gernsbacher, M.A., Pew, R.W., Hough, L.M., Pomerantz, J.R. (Eds.), Psychology and the Real World: Essays Illustrating Fundamental Contributions to Society. Worth Publishers, New York, pp. 56–64.

Brewer, S., Klein, J.D., 2006. Type of positive interdependence and affiliation motive in an asynchronous, collaborative learning environment. Educ. Technol. Res. Dev. 54 (4), 331–354.

Chapman, K.J., Van Auken, S., 2001. Creating positive group experiences: an examination of the role of the instructor on students' perceptions of group projects. J. Market. Educ. 23, 117–127.

Chapman, K.J., Meuter, M., Toy, D., Wright, L., 2006. Can't we pick our own groups? The influence of group selection method on group dynamics and outcomes. J. Manag. Educ. 30, 557–569.

Cooper, K.M., Downing, V.R., Brownell, S.E., 2018. Learning Anxiously: Alleviating and Exacerbating Student Anxiety in Active Learning Classrooms. Poster, SABER West.

Crouch, C.H., Mazur, E., 2001. Peer instruction: ten years of experience and results. Am. J. Phys. 69, 970–977.

Eddy, S.L., Brownell, S.E., Wenderoth, M.P., 2014. Gender gaps in achievement and participation in multiple introductory biology classrooms. CBE Life Sci. Educ. 13 (4), 738.

Eddy, S.L., Browness, S.E., Thummapha, P., Lan, M.C., Wenderoth, M.P., 2015. Caution, Student experience may vary: social identities impact a student's experience in peer discussions. CBE Life Sci. Educ. 14, 1–17.

Feichtner, S.B., Davis, E.A., 1984. Why some groups fail: a survey of students' experiences with learning groups. J. Manag. Educ. 9, 58–73.

Freeman, S., O'Connor, E., Parks, J.W., Cunningham, M., Hurley, D., Haak, D., et al., 2007. Prescribed active learning increases performance in introductory biology. CBE Life Sci. Educ. 6, 132–139.

Heller, P., Hollabaugh, M., 1992. Teaching problem solving through cooperative grouping. Part 2: Designing problems and structuring groups. Am. J. Phys. 60, 637–644.

James, M.C., 2006. The effect of grading incentive on student discourse in peer instruction. Am. J. Phys. 74, 689–691.

James, M.C., Barbieri, F., Garcia, P., 2008. What are they talking about? Lessons learned from a study of peer instruction. Astron. Educ. Rev. 7, 37.

Johnson, D.W., Johnson, R.T., 2009. An educational psychology success story: social interdependence theory and cooperative learning. Educ. Res. 38, 365–379.

Johnson, D.W., Johnson, R.T., Smith, K.A., 2007. The state of cooperative learning in postsecondary and professional settings. Educ. Psychol. Rev. 19, 15–29.

Johnson, D.W., Johnson, R.T., Smith, K.A., 2014. Cooperative learning: improving university instruction by basing practice on validated theory. J. Excell. Coll. Teach. 25, 85–118.

Kagan, S., 2014. Kagan structures, processing, and excellence in college teaching. J. Excell. Coll. Teach. 25, 119–138.

Kirschner, F., Paas, F., Kirschner, P.A., 2011. Task complexity as a driver for collaborative learning efficiency: the collective working-memory effect. Appl. Cogn. Psychol. 25, 615–624.

Knight, J.K., Wise, S.B., Southard, K.M., 2013. Understanding clicker discussions: student reasoning and the impact of instructional cues. CBE Life Sci. Educ. 12, 645–654.

Knight, J.K., Wise, S.B., Sieke, S., 2016. Group random call can positively affect student in-class clicker discussions. CBE Life Sci. Educ. 15 (4), ar56.

Kuh, G.D., Kinzie, J., Buckley, J., Bridges, B., Hayek, J.C., 2007. Piecing Together the Student Success Puzzle: Research, Propositions, and Recommendations. ASHE Higher Education Report, No. 32. Jossey-Bass, San Francisco, CA.

Lerner, L.D., 1995. Making student groups work. J. Manag. Educ. 19 (1), 123–125.

Lou, Y., Abrami, P.C., d'Apollonia, S., 2001. Small group and individual learning with technology: a meta-analysis. Rev. Educ. Res. 71, 449–521.

McDaniel, M.A., Butler, A.C., 2011. A contextual framework for understanding when difficulties are desirable. In: Benjamin, A.S. (Ed.), Successful Remembering and Successful Forgetting: A Festschrift in Honor of Robert A. Bjork. Psychology Press, New York, pp. 175–198.

Myers, S.A., 2012. Students' perceptions of classroom group work as a function of group member selection. Commun. Teach. 26, 50–64.

Scager, K., Boonstra, J., Peeters, T., Vulperhorst, J., Wiegant, F., 2016. Collaborative learning in higher education: evoking positive interdependence. CBE Life Sci. Educ. 15, 1–9.

Schmidt, H.G., Rotgans, J.I., Yew, E.H.J., 2011. The process of problem-based learning: what works and why. Med. Educ. 45, 792–806.

Sears, D.A., Pai, H.H., 2012. Effects of cooperative versus individual study on learning and motivation after reward-removal. J. Exp. Educ. 80 (3), 246–262.

Serrano, J.M., Pons, R.M., 2007. Cooperative learning: we can also do it without task structure. Intercult. Educ. 18 (3), 215–230.

Smith, M.K., Wood, W.B., Adams, W.K., Wieman, C., Knight, J.K., Guild, N., et al., 2009. Why peer discussion improves student performance on in-class concept questions. Science 323, 122–124.

Springer, L., Stanne, M.E., Donovan, S.S., 1999. Effects of small-group learning on undergraduates in science, mathematics, engineering, and technology: a meta-analysis. Rev. Educ. Res. 96 (1), 21–51.

Stowell, J.R., Nelson, J.M., 2007. Benefits of electronic audience response systems on student participation, learning, and emotion. Teach. Psychol. 34 (4), 253–258.

Treen, E., Atanasova, C., Pitt, L., Johnson, M., 2016. Evidence from a large sample on the effects of group size and decision-making time on performance in a marketing simulation game. J. Market. Educ. 38, 130–137.

Wenzel, T.J., 2007. Evaluation tools to guide students' peer-assessment and self-assessment in group activities for the lab and classroom. J. Chem. Educ. 84, 182–186.

Chapter 6

Metacognitive Practices: Giving Students Tools to be Self-Directed Learners

For many scientists, one of our main goals is for our students to become independent, self-directed learners. Many of the elements discussed in this book can contribute to this goal, from assignments that inspire and engage students to active learning approaches and group work that give them support while they tackle hard challenges. The key to helping students develop this self-direction is teaching them to be metacognitive—that is, to monitor and regulate their own learning. This chapter describes definitions for metacognition and reasons that we should foster it in our students. It also suggests specific practices that can be integrated into your course, becoming part of assignments, active learning exercises, group work, and even exams. This regular integration of metacognitive prompts into your course can have a powerful effect on your students' learning, both present and future.

WHAT IS IT?

Metacognition is very generally defined as thinking about one's own thinking. Since the term was coined in the 1970s by John Flavell, metacognition has come to be recognized as an important element for learning, allowing students to monitor and evaluate their understanding and choose strategies well suited to the learning task (Flavell, 1979; Pintrich, 2002). By many definitions, metacognition includes both knowledge of cognition and regulation of cognition (Schraw et al., 2006). That is, to be metacognitive, a student must have some knowledge of how learning works, perhaps including an ability to articulate processes that lead to memory formation, but certainly including knowledge of learning strategies and where and when they are likely to be effective. Importantly, however, a metacognitive student must also be able to regulate her own learning, engaging in processes for planning, monitoring, and evaluating.

Anyone reading this book is likely to be reflexively, unconsciously metacognitive, particularly in areas of greater expertise. As we learn more in a

Science Teaching Essentials. DOI: https://doi.org/10.1016/B978-0-12-814702-3.00006-8

domain, it becomes easier for us to assess what we do not know as well as for us to expand our understanding. These skills are not automatic, however, and they may be quite unfamiliar to our students (Pascarella and Terenzini, 2005). Further, ways of thinking and metacognitive skills are not identical across domains; the learning approaches that serve well in physics are not as well suited to biology, and still less so to literary analysis. Thus, teaching our students about the types of thinking that are prevalent in our discipline and the associated ways we monitor and evaluate our understanding can empower them to become more flexible, independent learners of the very content we care about.

WHY SHOULD WE CONSIDER OUR STUDENTS' METACOGNITION?

How People Learn, published by the National Research Council, synthesizes decades of cognitive science research to identify key features of effective learning environments (National Research Council, 2000). One of the central tenets of this report is that explicitly teaching students to be metacognitive within each content area promotes learning and improves students' ability to transfer knowledge to new settings (White and Frederiksen, 1998; Schoenfeld, 1991). The authors emphasize that metacognitive strategies are not generic, and that attempts to teach them as such can reduce students' ability to use them to transfer knowledge.

The importance of metacognition is echoed in *How Learning Works*, which identifies how seven principles of learning can inform college teaching (Ambrose et al., 2010). One principle states that, "To become self-directed learners, students must learn to assess the demands of the task, evaluate their own knowledge and skills, plan their approach, monitor their progress, and adjust their strategies as needed," describing the actions students must engage in to be metacognitive (p. 191). Novices within a field have a hard time with several of these actions. They tend to have limited understanding of effective learning strategies and may choose strategies that give them a greater sense of fluency and comfort with the material but that are less effective for long-term learning (Bjork et al., 2013; Dye and Stanton, 2017). They also tend to overestimate their current knowledge and underestimate the impact of future study on their ability to learn more, leading them to ineffective monitoring and planning for future study (Bjork et al., 2013). Further, they may fail to shift strategies when their initial approach does not work (Dye and Stanton, 2017; Fu and Gray, 2004). And, unsurprisingly, students who exhibit poor metacognitive skills are less successful academically (Kruger and Dunning, 1999; Dunning et al., 2003; May and Etkina, 2002).

However, as *How People Learn* articulated two decades ago, incorporating mechanisms to teach students metacognitive skills in our classes can lead to greater student learning (Bielaczyc et al., 1995; Chi et al., 1994;

Palinscar and Brown, 1984). This has been shown in in biology (Sabel et al., 2017), chemistry (Sandi-Urena et al., 2011; Cook et al., 2013), and physics (Taasoobshirazi et al., 2015), and it is widely supported in the geoscience and astronomy education literature (e.g., https://serc.carleton.edu/NAGTWorkshops/metacognition/index.html). By teaching these metacognitive skills in each of our classes, we help ensure that students learn these skills in ways that will help students learn to plan, monitor, and evaluate their learning *in our discipline*, developing knowledge that they can more readily transfer to new problems and settings.

HOW CAN WE PROMOTE OUR STUDENTS' METACOGNITION?

Helping students become more metacognitive involves both increasing their knowledge of cognition—that is, how we learn—and helping them develop skills to plan, monitor, and evaluate their understanding. The suggestions provided here range from general suggestions intended to increase students' understanding of how we learn to more specific strategies that can be incorporated into classes to support and scaffold self-monitoring and regulation.

1. *Teach students the basics of memory formation and associated strategies that promote learning.* Theoretical models of metacognition often break it down into two elements: knowledge of cognition and regulation of cognition (e.g., Schraw et al., 2006). Many of our students have very limited knowledge of cognition, including an understanding of how memories form and strategies that are effective at promoting memory formation. By introducing a model of memory (such as the one presented in Chapter 4) and strategies for learning that have been demonstrated to promote learning, you can empower students to make more informed choices about their study approaches. Which study strategies should you share? Students should know that self-testing and distributed (or spaced) practice are highly effective study strategies that appear to work across contexts and populations (Dunlosky et al., 2013; Bjork et al., 2013). Further, self-explanation and interleaved practice—that is, mixing up study rather than massing practice at a single skill or content area—also have strong support. Further, it may be helpful for students to know that these strategies are less comfortable than other common but less effective techniques, such as rereading, highlighting, and summarizing. How you teach this knowledge of cognition will vary depending on the teacher, the students, and other elements of the course, but it will be very helpful to ask students to consider how these ideas resonate with and connect to their own experience, helping provide a hook that may increase the likelihood that students will remember and value these approaches.

2. *Teach students about different types of knowledge.* Another area in which students' understanding of cognition can be expanded involves the types of knowledge that are the targets for their learning. It can be particularly useful for students' metacognition to distinguish between declarative and procedural knowledge (Ambrose et al., 2010; Oosterhof, n.d.). Declarative knowledge is knowing about something and being able to describe it with words or symbols, either verbally or in writing. Procedural knowledge, in contrast, is knowing how and when to do something, and involves applying concepts and rules that describe relationships. Both declarative knowledge and procedural knowledge can be quite complex and hard to achieve. For example, explaining why concepts are related or how something works demonstrates declarative knowledge, whereas categorizing problems according to the principle used to solve them demonstrates procedural knowledge, and both of these can be tasks that challenge students. The types of practice that can help our students achieve these goals differ, however, and one step toward helping students appropriately plan and evaluate their learning is for them to recognize the differences in the types of knowledge they are pursuing. Students may inaccurately think that being able to recognize or describe a phenomenon means that they are able to apply it in a problem, or, alternatively, they may think that being able to solve a problem means that they are able to explain a concept. By differentiating between these two types of knowledge and encouraging students to ask themselves which they should be able to demonstrate—and monitor and evaluate whether they can, in fact, do that—instructors can provide students with additional metacognitive tools (Ambrose et al., 2010).

3. *Provide students prompts that help them plan, monitor, and evaluate their learning.* Tanner's (2012) excellent essay on promoting student metacognition suggests questions that students can ask themselves to help plan, monitor, and evaluate their approaches to several different learning experiences, from a class session to a homework assignment to the entire course. For example, she suggests students ask themselves the following questions about a homework assignment:

Planning	What is the instructor's goal in having me do this task? What are all the things I need to do successfully to accomplish this task? How much time do I need? If I have done something like this before, how could I do a better job this time?
Monitoring	What strategies am I using that are working well or not working well? What other resources could I be using? What is most challenging/most confusing for me about this task?
Evaluating	To what extent did I successfully accomplish the goals of the task? If I were the instructor, what would I identify as strengths of my work and flaws in my work? When I do an assignment like this again, what do I want to remember to do differently?

Tanner suggests that questions like this can be shared directly with students or embedded into assignments.

4. *Incorporate metacognitive prompts into course dialog.* Tanner also suggests that building prompts that promote metacognitive activity into the classroom conversation can normalize metacognitive practices and make students come to think of them as part of normal studying process (Tanner, 2012; May and Etkina, 2002). Asking students to explain their reasoning to a partner during a clicker question discussion prompts students to monitor their thinking, and asking them to indicate how confident they are in their answer and to identify what they would need to know to increase their confidence encourages them to evaluate and to plan. Further, modeling your metacognitive processes can give students other examples of ways to approach problems. For example, you could "think aloud" through a homework problem or writing assignment, talking about how you start the process, how you would proceed, and how you would check yourself along the way (Ambrose et al., 2010). Importantly, this gives you a chance to demonstrate how you use approaches for quickly assessing your work, such as asking if an answer is reasonable or uses relevant assumptions.

5. *Use the 3−2−1.* The 3−2−1 is a reading response technique that that is extraordinarily easy to use and adaptable to different disciplines, class settings, and even assignments (i.e., it can be used for more than reading response). In the classic version of the 3−2−1, students are asked to identify the three most important ideas from their reading, two things that they did not understand, and one question that they would like to ask the author that goes beyond the reading (Van Gyn, 2013). By asking students to identify the most important ideas and areas of confusion, it provides tools that help them monitor their understanding, and by asking them to identify a question that goes beyond the reading, it arguably prompts students to begin to plan their next steps. Although the 3−2−1 was originally described as a reading response technique, it can readily be applied to review of videos, class sessions, or even problem-solving. The broad usefulness of the 3−2−1, however, really lies in the mutability of the "1." Rather than asking students to identify a question that goes beyond the reading, instructors can make the "1" be any prompt that is useful for their class. For example, you might ask your students to write an exam question over the material, sketch a figure that summarizes one part of the reading, or predict the next step in a process. By talking to the students about how these steps allow them to monitor and evaluate their understanding, you give students a set of metacognitive tools that you and they can apply easily and broadly to course material.

6. *Use exam wrappers.* Exam wrappers provide students with a structured set of questions to help them reflect on their exam performance and

how it relates to their exam preparation (Lovett, 2013). When an exam is returned, it is "wrapped" with a cover sheet with questions that ask students to consider how and when they studied, how they did on the exam, and what their plan should be for studying for the next exam. Students complete the wrapper in class or as homework and turn it in, and instructors return it to them a couple of weeks before the next exam. The process has been shown to improve student performance in some cases but not in others (Gizen Gezer-Templeton et al., 2017; Soicher and Gurung, 2017). The instructor's introduction and consistent use of the tool may impact students' response, but it is also true that students do not always adopt good approaches they have identified for themselves but instead revert to familiar and comfortable patterns (Dye and Stanton, 2017). Nonetheless, the exam wrapper gives students a tool to analyze and connect their study and exam performance in a way that can help them plan for future success. Note that the wrapper can be adapted to other kinds of assignments that students do repeatedly in a course, such as lab reports or papers.

7. *Use enhanced answer keys and reflection questions.* In large, introductory science courses, instructors often find it difficult to provide feedback to students with a frequency that enables them to adequately monitor their progress and understanding. In an attempt to combat this challenge, Sabel et al. (2017) investigated the use of enhanced answer keys with reflection questions to determine whether these self-assessment tools can be used to provoke self-regulatory learning processes. The enhanced answer keys were provided for both homework assignments and exams and included explanations related to common errors as well as reflection questions such as, "Do all of the answers on the answer key make sense to you? If not, what steps can you take to make sense of the concepts in the answers? What could you have done differently before or while you completed this assignment to understand better this topic? What are some possible questions you might see on a future exam about this topic?" They found that students who engaged with the enhanced answer keys exhibited some form of metacognition and performed better on assignments and in the course than those who did not use them. They also found, however, that explicit instruction on how to use the keys was needed for some students. Thus, enhanced answer keys that include reflection questions may provide a low-stakes way to incorporate metacognitive prompts into your course.

8. *Consider a problem-solving exercise designed to develop metacognitive skills.* Sandi-Urena et al. evaluated the impact of an intervention in which groups of two or three students collaborated to solve a problem that appears complex but, in reality, is very simple (e.g., what is the minimum number of socks needed to have two of the same color when there is a ratio of 4 brown to 5 black socks?). The problem was

designed to generate a cognitive imbalance experience that triggers reflection, with support for reflection provided by 26 explicitly metacognitive prompts that the students responded to within their groups. The students then completed 13 additional prompts individually as homework. The final stage of the exercise involved instructors providing a summary of the findings intended to reinforce students' awareness of metacognitive skills and was accompanied by eight prompts to which students responded individually. The initial problem-solving session lasted approximately 45 minutes and the final summary session required about 10 additional minutes in class. This treatment increased students' ability to solve ill-structured problems requiring significant metacognitive monitoring (Sandi-Urena et al., 2011). This study was performed in a General Chemistry laboratory, but because the initial problems do not require particular knowledge of chemistry, the approach may also prove fruitful in other disciplines.

9. *Use the Model−Observe−Reflect−Explain (MORE) framework to guide lab work.* The MORE thinking process asks students to consider their initial understanding of a system (Model), conduct experiments to test that understanding (Observe), consider what their observations mean (Reflect), and revise their initial understanding (Explain) (Mattox et al., 2006). In the Model portion of the process, which is assigned as homework and then reviewed in small groups in lab, students are asked to describe their understanding of what will happen in a physical process. For example, they may be asked to describe what will happen when salt and sugar are placed in water, providing both a description of what they will see and a drawing of what will happen at the microscopic level. They then perform an experiment and analyze the results, looking for patterns or trends, and consider how these results relate to their initial models. Finally, they are asked to share their data with the class and make any needed modifications to their initial models that allow the models to explain the data. The approach has primarily been used in chemistry labs but has the potential to be adapted to other disciplines as well, and its structure incorporates multiple metacognitive prompts that can help students think about their understanding and how it can change in response to new information. Further, the use of this framework has been associated with positive changes in student attitudes (Schroeder et al., 2018).

10. *Consider adopting self-assessment checklists that you use throughout a semester.* The American Council on the Teaching of Foreign Languages has identified a series of "can-do statements" that are performance indicators for language learners (NCSSFL-ACTFL, 2015). These statements include benchmarks for interpersonal communication (e.g., "I can communicate and exchange information about familiar topics using phrases and simple sentences..."), presentational speaking, interpretive

listening, and other skills that students can use to identify progress throughout a course or series of courses. Developing a set of "can-do" statements for your course may provide students both a series of guide-posts about where they are going as well as a tool to evaluate how they are progressing. Alternatively, some instructors use knowledge surveys to serve the same functions. Knowledge surveys cover the full content of a course and consist of questions or problems that students should learn how to address during the course (Wirth and Perkins, 2005). Rather than answering the questions, however, students indicate how confident they are that they could answer the question, choosing responses like "I do not understand the question," "I am not confident that I could answer the question well enough for grading purposes at this time," or "I understand the question and am confident I could find the information I need to answer it in less than 20 minutes." Because students do not answer the questions but indicate their confidence about them, knowledge surveys can cover a broad range of questions in a rela-tively short amount of time. Both of these tools help students under-stand the goals for their learning and give them a low-stakes way to assess their progress.

CONCLUSION

There are many other excellent resources for learning more about metacogni-tion and how to promote its development in your students. Chapter 7 in *How Learning Works* provides a helpful framework and associated teaching strate-gies for helping students become self-directed learners, and both Linda Hodges' *Teaching Undergraduate Science* and Linda Nilson's *Creating Self-Regulated Learners: Strategies to Strengthen Students' Self-Awareness and Learning Skills* are highly readable, informative and practical books that focus explicitly on ways to promote student metacognition and self-regulation. Like many areas of teaching, figuring out how effectively to pro-mote student metacognition can be a lifelong pursuit, but the key to getting started is to be aware that we should be helping our students develop their metacognitive skills and thoughtful about adopting an approach that works for our course and our students.

One key element of metacognition is knowledge of effective learning techniques, and the next chapter focuses on an approach that has been repeat-edly shown to promote learning. Variously called the testing effect and test-enhanced learning, retrieval practice has been shown to enhance memory and understanding in a variety of contexts. It is a fourth important, adaptable teaching approach that can be part of and can complement the active learning, group work, and metacognitive practices that we use to help our students learn.

REFERENCES

Ambrose, S.A., Bridges, M.W., DiPietro, M., Lovett, M.C., Norman, M.K., 2010. How Learning Works: Seven Research-Based Principles for Smart Teaching. Jossey-Bass, San Francisco, CA.

Bielaczyc, K., Pirolli, P.L., Brown, A.L., 1995. Training in self-explanation and self-regulation strategies: investigating the effects of knowledge acquisition activities on problem solving. Cogn. Instruct. 13, 221−252.

Bjork, R.A., Dunlosky, J., Kornell, N., 2013. Self-regulated learning: beliefs, techniques, and illusions. Annu. Rev. Psychol. 64, 417−444.

Chi, M.T.H., deLeeuw, N., Chiu, M., LaVancher, C., 1994. Eliciting self-explanations improves understanding. Cogn. Sci. 18, 439−477.

Cook, E., Kennedy, E., McGuire, S.Y., 2013. Effect of teaching metacognitive learning strategies on performance in general chemistry courses. J. Chem. Educ. 90, 961−967.

Dunlosky, J., Rawson, K.A., Marsh, E.J., Nathan, M.J., Willingham, D.T., 2013. Improving students' learning with effective learning techniques: promising directions from cognitive and educational psychology. Psychol. Sci. Public Interest 14, 4−58.

Dunning, D., Johnson, K., Ehrlinger, J., Kruger, J., 2003. Why people fail to recognize their own incompetence. Curr. Dir. Psychol. Sci. 12, 83−87.

Dye, K.M., Stanton, J.D., 2017. Metacognition in upper-division biology students: awareness does not always lead to control. CBE Life Sci. Educ. 16, ar31. Available from: https://doi.org/10.1187/cbe.16-09-0286.

Flavell, J., 1979. Metacognition and cognitive monitoring: a new area of cognitive-developmental inquiry. Am. Psychol. 34, 906−911.

Fu, W.-T., Gray, W.D., 2004. Resolving the paradox of the active user: stable suboptimal performance in interactive tasks. Cogn. Sci. 28 (6), 901−935.

Gizen Gezer-Templeton, P., Mayhew, E.J., Korte, D.S., Schmidt, S.J., 2017. Use of exam wrappers to enhance students' metacognition n a large introductory food science and human nutrition course. J. Food Sci. Educ. 16, 28−36.

Kruger, J., Dunning, D., 1999. Unskilled and unaware of it: how difficulties in recognizing one's own incompetence lead to inflated self-assessments. J. Pers. Soc. Psychol. 77 (6), 1121−1134.

Lovett, M.C., 2013. Make exams worth more than the grade. In: Kaplan, M., others (Eds.), Using Reflection and Metacognition to Improve Student Learning: Across the Disciplines, Across the Academy. Stylus Publishing, Sterling, VA, pp. 18−52.

Mattox, A.C., Reisner, B.A., Rickey, D., 2006. What happens when chemical compounds are added to water? An introduction to the Model−Observe−Reflect−Explain (MORE) thinking frame. J. Chem. Educ. 83, 622−624.

May, D.B., Etkina, E., 2002. College physics students' epistemological self-reflection and its relationship to conceptual learning. Am. J. Phys. 70, 1249−1258.

National Research Council, 2000. How People Learn: Brain, Mind, Experience, and School: Expanded Edition. The National Academies Press, Washington, DC. Available from: https://doi.org/10.17226/9853.

NCSSFL-ACTFL, 2015. Can-do statements, 2015. <https://www.actfl.org/sites/default/files/pdfs/Can-Do_Statements_2015.pdf>.

Oosterhof, A., n.d. How different types of knowledge are assessed. Retrieved from: <http://www.cala.fsu.edu/modules/assessing_knowledge/> (accessed 22.03.18.).

Palinscar, A.S., Brown, A.L., 1984. Reciprocal teaching of comprehension monitoring activities. Cogn. Instr. 1, 117−175.

Pascarella, E.T., Terenzini, P.I., 2005. How College Affects Students: A Third Decade of Research. Jossey-Bass, San Francisco, CA.

Pintrich, P.R., 2002. The role of metacognitive knowledge in learning, teaching, and assessing. Theory Pract. 41 (4), 219−225.

Sabel, J.L., Dauer, J.T., Forbes, C.T., 2017. Introductory biology students' use of enhanced answer keys and reflection questions to engage in metacognition and enhance understanding. CBE Life Sci. Educ. 16 (3), 1−12. ar40.

Sandi-Urena, S., Cooper, M.M., Stevens, R.H., 2011. Enhancement of metacognition use and awareness by means of a collaborative intervention. Int. J. Sci. Educ. 33, 323−340.

Schoenfeld, A.H., 1991. On mathematics as sense-making: an informal attack on the unfortunate divorce of formal and informal mathematics. In: Voss, J.F., Perkins, D.N., Segal, J.W. (Eds.), Informal Reasoning and Education. Erlbaum, Hillsdale, NJ, pp. 331−343.

Schraw, T., Crippen, K.J., Hartley, K., 2006. Promoting self-regulation in science education: metacognition as part of a broader perspective on learning. Res. Sci. Educ. 36, 111−139.

Schroeder, L., Bierdz, J., Wink, D.J., King, M., Daubenmire, P.L., Clark, G.A., 2018. Relating chemistry to healthcare and MORE: implementation of MORE in a survey organic and biochemistry course for prehealth students. J. Chem. Educ. 95, 37−46.

SERC, Understanding what our geoscience students are learning: observing and assessing. <https://serc.carleton.edu/NAGTWorkshops/assess/knowledgesurvey/index.html>.

Soicher, R.N., Gurung, R.A.R., 2017. Do exam wrappers increase metacognition and performance? A single course intervention. Psychol. Learn. Teach. 16 (1), 64−73.

Taasoobshirazi, G., Bailey, M.L., Farley, J., 2015. Physics metacognition inventory part II: Confirmatory factor analysis and Rasch analysis. Int. J. Sci. Educ. 37, 2769−2786.

Tanner, K.D., 2012. Promoting student metacognition. CBE Life Sci. Educ. 11, 113−120.

Van Gyn, G., 2013. The little assignment with the big impact: reading, writing, critical reflection, and meaningful discussion. Faculty Focus. <https://www.facultyfocus.com/articles/instructional-design/the-little-assignment-with-the-big-impact-reading-writing-critical-reflection-and-meaningful-discussion/>.

White, B.Y., Frederiksen, J.R., 1998. Inquiry, modeling, and metacognition: making science accessible to all students. Cogn. Instr. 16, 3−118.

Wirth, K.R., Perkins, D., 2005. Knowledge surveys: an indispensable course design and assessment tool. Presented at the Innovations in the Scholarship of Teaching and Learning in the Liberal Arts at St. Olaf College. Retrieved from: <https://www.macalester.edu/academics/geology/wirth/WirthPerkinsKS.pdf> (accessed 22.03.18.).

Chapter 7

Test-Enhanced Learning: Using Retrieval Practice To Help Students Learn[1]

with Rachel E. Biel

Almost all science classes incorporate testing, often as summative assessments at the end of a course segment, formative assessments that measure progress and provide feedback along the way, or diagnostic tools at the beginning of a course. Rarely, however, do we think of tests as learning tools. We may acknowledge that testing promotes student learning, but we often attribute this effect to the studying that students do to prepare for the test. And yet, one of the most consistent findings in cognitive psychology is that testing is more effective than studying at promoting learning (Roediger and Butler, 2011; Roediger and Pyc, 2012). Given the potential power of testing as a tool to promote learning, we should consider how to incorporate testing into our courses not only to gauge students' learning, but also to promote learning (Klionsky, 2008). This chapter defines test-enhanced learning, describes evidence that it can promote learning, and provides suggestions for ways to incorporate testing-for-learning in your course.

WHAT IS IT?

Test-enhanced learning is the idea that the process of remembering concepts or facts—retrieving them from memory—increases long-term retention of those concepts or facts. This idea is also known as the testing effect and derives from studies examining the effect of various types of tests, or prompts, that promote retrieval of information from long-term memory. In some ways, the terms test-enhanced learning and the testing effect are misnomers, in that the use of the word "tests" typically calls up notions of

---

1. *Note:* This chapter is modified from an article first published in *CBE—Life Sciences Education. Brame, C.J., & Biel, R. (2015). Test-enhanced learning: the potential for testing to promote greater learning in undergraduate science courses. CBE—Life Sci Educ 14: es4.*

Science Teaching Essentials. DOI: https://doi.org/10.1016/B978-0-12-814702-3.00007-X

high-stakes summative assessments. In fact, most or all studies elucidating the testing effect examine the impact of low-stakes retrieval practice on a delayed summative assessment. The "testing" that actually enhances learning is the low-stakes retrieval practice that accompanies study in these experiments. With that caveat in mind, the testing effect can be a powerful tool to add to both instructors' and students' tool kits.

WHAT DO WE KNOW ABOUT THE EFFECTS OF TESTING?

1. *Repeated retrieval enhances long-term retention in a laboratory setting.* The idea that active retrieval of information from memory improves memory is not a new one: William James proposed this idea in 1890, and Edwina Abbott and Arthur Gates provided support for this idea in the early part of the 20th century (James, 1890; Abbott, 1909; Gates, 1917). During the last decade, however, evidence of the benefits of testing has mounted. Here, we summarize two studies illustrating this effect in undergraduates learning educationally relevant materials in a laboratory setting.

   Roediger and Karpicke investigated the effects of single versus multiple testing events on long-term retention using educationally relevant conditions (Roediger and Karpicke, 2006a). Their goal was to determine if any connection existed between the number of times students were tested and the size of the testing effect. The investigators worked with undergraduates in a laboratory environment, asking them to read passages about 250 words long. The authors compared three conditions: students who studied the passages four times for 5 minutes each (SSSS group); students who studied the passages three times and completed one recall test in which they were given a blank sheet of paper and asked to recall as much of the passage as they could (SSST group); and students who studied the passages one time and then performed the recall practice three times (STTT group). Student retention was then tested either 5 minutes or 1 week later using the same type of recall test used for retrieval practice. Interestingly, results differed significantly depending on when the final test was performed. Students who took their final test very soon after their study period (i.e., 5 minutes) benefited from repeated studying, with the SSSS group performing best, the SSST group performing second-best, and the STTT group performing least well. This result suggests that studying is more effective when the information being learned is only needed for a short time—a result that may not be surprising to anyone who has crammed for a test. However, the results were reversed when the final test was delayed by a week, with the STTT group performing about 5% higher than the SSST group and about 21% higher than the SSSS group. Testing had a greater impact on long-term retention than did repeated study, and the participants who were repeatedly tested had increased retention over those who were only tested once.

Karpicke and Blunt sought to evaluate the impact of retrieval practice on students' learning of undergraduate-level science concepts, comparing the effects of retrieval practice to the elaborative study technique, concept mapping (Karpicke and Blunt, 2011). In one experiment, students studied a science text and were then divided into one of four conditions: a study-once condition, in which they did not interact further with the concepts in the text; a repeated study condition, in which they studied the text four additional times; an elaborative study condition, in which they studied the text one additional time, were trained on concept mapping, and produced a concept map of the concepts in the text; and a retrieval practice condition, in which they completed a free recall test, followed by an additional study period and recall test. All students were asked to complete a self-assessment predicting their recall within 1 week; students in the repeated study group predicted better recall than students in any of the other groups. Students then returned a week later for a short-answer test consisting of questions that could be answered verbatim from the text and questions that required inferences from the text. Students in the retrieval practice condition performed significantly better on both the verbatim questions and the inference questions than students in any other group. The authors did additional experiments investigating whether the results would differ for specific students and whether the advantage of retrieval practice would persist if the final test consisted of a concept mapping exercise. The authors observed that retrieval practice produced better performance on both types of final tests (short-answer and concept mapping). Further, when they examined the effects on individual learners, they found that 84% (101/120) students performed better on the final tests when they used retrieval practice as a study strategy rather than concept mapping.

In short, these studies suggest that retrieving information from long-term memory is highly effective for promoting learning when compared with other common study strategies. These summaries provide only an introduction to the rich literature on the testing effect; several review articles provide a thorough overview of the work in this area (Roediger and Butler, 2011; Roediger and Karpicke, 2006b; Roediger et al., 2011).

2. *Various testing formats can enhance learning.* The studies described above used free recall as a "testing" strategy, but additional studies have investigated whether other testing formats can also be effective. Smith and Karpicke examined the effects of different types of testing questions in a series of experiments with undergraduates in a laboratory environment (Smith and Karpicke, 2014). In one experiment, five groups of students were compared. Students read four texts, each approximately 500 words long. After each, four groups of students then participated in

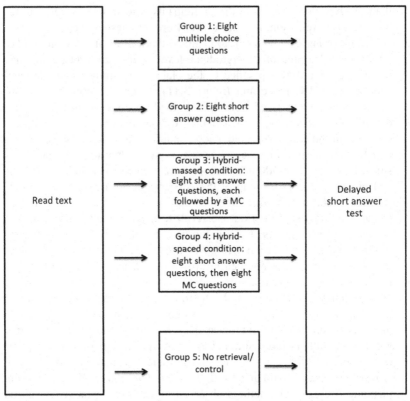

FIGURE 7.1   Experimental design of Smith and Karpicke (2014) experiment examining effect of question format on testing effect.

different types of retrieval practice, while the fifth group was the no-retrieval control (see Fig. 7.1). 1 week later, the students returned to the lab for a short-answer test on each of the reading passages. Confirming other studies, students who had participated in some type of testing performed much better on the final assessment, getting approximately twice as many questions correct as those who did not have any retrieval practice. This was true both for questions that were directly taken from information in the texts as well as questions that required inference from the text. There was no significant difference in the benefits conferred by the different types of testing; multiple-choice, short-answer, and a mix of both types of questions following the reading were equally effective at enhancing the students' learning. Other experiments in the series essentially replicated these results.

Kang, McDermott, and Roediger also compared the effects of multiple-choice questions and short-answer questions on undergraduate students' ability to recall information from short articles after a 3-day delay (Kang et al., 2007). In the final test, they used both short-answer and multiple-choice questions. They observed that when students answered either short-answer or multiple-choice questions after reading the article, they recalled more information on the final test, whether the questions on the final test were multiple-choice or short-answer. When feedback (that is, the correct answer to the question) was given on the postreading test, short-answer questions provided slightly more benefit than did multiple-choice questions. However, when feedback was not provided, initial multiple-choice questions provided the greater benefit.

Together, these and other studies suggest that multiple question formats can provide the benefit associated with testing.

3. *Feedback enhances the benefits of testing.* Considerable work has been done to examine the role of feedback on the testing effect. Butler and Roediger designed an experiment in which undergraduates studied 12 historical passages and then took multiple-choice tests in a lab setting (Butler and Roediger, 2008). The students either received no feedback, immediate feedback (i.e., following each question), or delayed feedback (i.e., following completion of the 42-item test). 1 week later, the students returned for a comprehensive test. While simply completing multiple-choice questions after reading the passages did improve performance on the final test, feedback provided an additional benefit, with delayed feedback resulting in better final performance than immediate feedback.

In a follow-up study, Butler, Karpicke, and Roediger demonstrated that feedback can provide a particular benefit by strengthening student recall of correct but low-confidence responses (Butler et al., 2008). Working with undergraduates in a laboratory setting, they asked students multiple-choice items about general knowledge (e.g., *What is the longest river in the world?*), following each item with a prompt to determine confidence in the answer (i.e., 1 = guess, 4 = high confidence). Students then received feedback for some of the multiple-choice items but no feedback for others. After a 5-minute delay, students completed a cued-recall test. While a testing effect was observed even in the absence of feedback, feedback strongly improved final performance, approximately doubling student performance over testing without feedback. This result was true both for questions that students had answered correctly and questions they had answered incorrectly on the initial multiple-choice test, but was most pronounced for low-confidence correct answers.

Thus, feedback on both low-confidence correct answers and incorrect answers may further enhance the testing effect, allowing students to solidify their understanding of concepts about which they are unclear.

4. *Learning is not limited to rote memory.* One concern that instructors may have about testing as a teaching and learning strategy is that it may promote rote memory. While most instructors recognize that memory plays

a role in allowing students to perform well within their academic domain, they want their students to be able to do more than simply remember and understand facts, but instead to achieve higher cognitive outcomes. However, several studies examined this question and report results suggesting that testing provides benefits beyond improving simple recall. For example, the studies by Karpicke and Blunt (2011) and Smith and Karpicke (2014) described above examined students' ability to answer questions that required inference, defined as drawing conclusions that were not directly stated within the passages but that could be drawn by synthesizing from multiple facts within the passage. In both studies, the investigators observed that testing following reading improved students' ability to answer inference questions on a delayed test.

Butler also examined whether test-enhanced learning can be used to promote transfer, or the ability to use facts and concepts from one domain in a different knowledge domain (Butler, 2010). He compared whether a restudy condition or an initial test was more effective at promoting students' ability to answer a question requiring transfer of knowledge, providing the following example:

The following concept was tested on the initial test: "A bat has a very different wing structure from a bird. What is the wing structure of a bat like relative to that of a bird?" (Answer: "A bird's wing has fairly rigid bone structure that is efficient at providing lift, whereas a bat has a much more flexible wing structure that allows for greater maneuverability.") The related inferential question about a different domain was the following: "The U.S. Military is looking at bat wings for inspiration in developing a new type of aircraft. How would this new type of aircraft differ from traditional aircraft like fighter jets?" (Answer: "Traditional aircrafts are modeled after bird wings, which are rigid and good for providing lift. Bat wings are more flexible, and thus an aircraft modeled on bat wings would have greater maneuverability.")

He observed that repeated testing improved students' ability to transfer knowledge to different domains when compared to restudy. Butler draws on the work of others to provide potential explanations for this effect, suggesting that retrieving information from memory may lead to the elaboration of existing retrieval routes or the development of additional retrieval routes (Bjork, 1975; McDaniel and Masson, 1985, within Butler, 2010).

5. *Testing potentiates further study.* Wissman, Rawson, and Pyc have reported work that suggests that retrieval practice over one set of material may facilitate learning of later material (Wissman et al., 2011). Undergraduate students were asked to read three sections of a text. In the "interim test" group, they were tested after reading each of the first two sections, specifically by typing everything they could remember about the text. After completing the interim test, they were advanced to the

next section of material. The "no interim test" group read all three sections with no tests in between. Both groups were tested on Section 3 after reading it. Interestingly, the group that had completed interim tests on Sections 1 and 2 recalled about twice as many "idea units" from Section 3 as the students who did not take interim tests. This result was observed both when Sections 1, 2, and 3 were about different topics and when they were about related topics. Thus, testing may help students learn material that they encounter after the testing event as well as the material for which they practice recall.

6. *The benefits of testing extend to the classroom.* All of the reports described above focused on experiments performed in a laboratory setting. In addition, there are several studies indicating that the benefits of testing extend to the classroom. The four summarized here were chosen because they illustrate different classroom-friendly approaches or benefits for a particular student population.

Larsen, Butler, and Roediger asked whether a testing effect was observed for medical residents' learning about status epilepticus and myasthenia gravis, two neurological disorders, at a didactic conference (Larsen et al., 2009). Specifically, residents participated in an interactive teaching session on the two topics and then were randomly divided into two groups. One group studied a review sheet on myasthenia gravis and took a test on status epilepticus, while the other group took a test on myasthenia gravis and studied a review sheet on status epilepticus. Six months later, the residents completed a test on both topics. The authors observed that the testing condition produced final test scores that averaged 13% higher than the study condition. This study demonstrates that testing can even provide benefits for a highly motivated, successful student population after a 6 month delay.

Lyle and Crawford examined the effects of retrieval practice on student learning in undergraduate statistics class (Lyle and Crawford, 2011). In one section of the course, students were instructed to spend the final 5−10 minutes of each class period answering two to four questions that required them to retrieve information about the day's lecture from memory. The students in this section of the course performed about 8% higher on exams over the course of the semester than students in sections that did not use the retrieval practice method, a statistically significant difference. This study demonstrates a very easy but effective way to incorporate retrieval practice and demonstrates benefit in a college-level science course.

McDaniel, Wildman, and Anderson examined the effects of unsupervised online quizzing on student performance in a web-based undergraduate Brain and Behavior class (McDaniel et al., 2012). The goal was to compare the effects of multiple-choice quiz questions, short-answer quiz questions, and targeted study of facts on students' performance on a unit exam corresponding to about 3 weeks' worth of class material. Students read textbook chapters weekly and took weekly online quizzes, which were graded for completion only. Some facts were targeted on the online quiz through multiple-choice questions, some were targeted through short-answer questions, some were targeted through representation of the fact, and some were not represented on the weekly quizzes. The authors observed that both types of quiz questions improved student performance on the unit exam in comparison to facts that were not targeted on the weekly quizzes, both on questions that were identical on the unit exam and the weekly quizzes and on questions that were related but not identical. Rereading the fact on the quiz only helped if the final exam question was identical to the presented fact. This study shows that simple online quizzes that are autograded for completion can improve student learning in a college science class.

Orr and Foster (2013) did a similar study in an introductory biology course for majors, examining the effects of frequent quizzing on student test performance. Using the Mastering Biology platform, Orr and Foster assigned 10 quizzes to students over the course of the semester. They then compared exam performance of students who took all or none of the quizzes, finding that students who took all of the quizzes performed significantly better than those who took none of the quizzes. Importantly, this trend was observed both for high-, middle-, and low-performing students, suggesting that frequent quizzing can provide benefit for students across a range of academic abilities.

WHAT ARE COMMON FEATURES OF "TESTS" THAT PROMOTE TEST-ENHANCED LEARNING?

The term "testing" evokes a certain response from most of us: the person being tested is being evaluated on his or her knowledge or understanding of a particular area, and will be judged right or wrong, adequate or inadequate based on the performance given. This implicit definition does not reflect the settings in which the benefits of test-enhanced learning have been established. In the experiments done in cognitive science laboratories, the "testing" was simply a learning activity for the students; in the language of the classroom, it could be considered a no-stakes formative assessment where students could evaluate their memory of a particular subject. In most of the studies from classrooms, the testing was either no-stakes recall practice (Larsen et al., 2009; Lyle and Crawford, 2011; Stanger-Hall et al., 2011) or low-stakes quizzes (McDaniel et al., 2012; Orr and Foster, 2013). Thus, the term retrieval practice may be a more accurate description of the activity that promoted students' learning. Implementing approaches to test-enhanced learning in a class should therefore involve no-stakes or low-stakes scenarios in which students are engaged in a recall activity to promote their learning rather than being repeatedly subjected to high-stakes testing situations. The distinction between high-stakes and low-/no-stakes "testing" is particularly important because of the consequences that high-stakes evaluation scenarios can have on identity-threatened groups. Stereotype threat is a phenomenon in which individuals in a stereotyped group underperform on high-stakes evaluations (Steele, 2010). In essence, social cues that activate a stereotype in an identity-threatened group generate anxiety about fulfilling the stereotype, producing a cognitive load that significantly impedes performance on the assessment. This phenomenon has been demonstrated for women in math, African Americans in higher education, white males in sports competitions, and a variety of other groups that are negatively stereotyped in a particular domain. Importantly, the effect is particularly potent in high-achieving individuals within that domain—for example, women who are high achievers in math are more likely to see a decline in their test performance if they are reminded of the stereotype that women are not good at math (Steele, 2010). In science classrooms, it may therefore be particularly important to consider approaches to test-enhanced learning that are no or low stakes and are articulated as learning opportunities, therefore minimizing the potential for stereotype threat.

WHAT ARE OPPORTUNITIES FOR IMPLEMENTATION IN THE CLASSROOM?

These results point to several possible implementations within the classroom.

- *Incorporating frequent quizzes.* These quizzes can consist of short-answer or multiple-choice questions, and can be administered online or face-to-face. The studies summarized above suggest that providing students the opportunity for retrieval practice—and ideally, providing feedback for the responses—will increase learning of targeted as well as related material.
- *Providing pauses during class.* At each of these pause points, students are asked either to recall the most important elements of the preceding class segment or to respond to instructor-generated questions that stimulate recall. This approach is similar to the pause procedure that is perhaps the simplest active learning technique (see Chapter 4), but asks students to remember information rather than to review notes. Importantly, the recall process can be beneficial when students simply think of but don't explicitly state a response as well as when they write or verbalize responses (Pyc et al., 2014). Importantly, this helps solidify memory not only of the material that students retrieve, but also improves their learning of material that follows (Wissman et al., 2011).

- *Incorporating retrieval practice at the end of class.* Setting aside the last few minutes of a class to ask students to recall, articulate, and organize their memory of the content of the day's class may provide significant benefits to their later memory of these topics. This approach is one adaptation of the minute paper active learning approach described in Chapter 4.

- *Using pretests.* All the studies reported in this chapter examine the effects of testing (or retrieval practice) after a subject has been introduced through study or classroom activities. There are other studies, however, that demonstrate that the act of taking a pretest can also prime students for learning (e.g., Pazicni and Pyburn, 2014; Little and Bjork, 2012; Richland et al., 2009). Thus, by giving students a no-stakes pretest before a unit or even a day of instruction, an instructor may help alert students both to the types of questions that they need to be able to answer as well as the key concepts and facts they need to be look for during study and instruction.

- *Telling students about the testing effect.* Finally, instructors may be able to aid their students' metacognitive abilities by sharing a synopsis of these observations. Telling students that frequent quizzing helps learning—and that effective quizzing can take a variety of forms—can give them a particularly helpful tool to add to their learning toolkit (Stanger-Hall et al., 2011). Adding the potential benefits of pretesting may further empower students to take control of their own learning, such as by using example exams as primers for their learning rather than simply as pre-exam checks on their knowledge.

It's important to note that incorporating testing—or recall practice—as a learning tool in a class should be done in conjunction with other evidence-based teaching practices described throughout this book. When considered through that lens, using retrieval practice as a learning tool may be a particularly valuable opportunity both to strengthen memory and to promote students' metacognition.

CONCLUSION

Test-enhanced learning is an underutilized, highly effective, and adaptable means to help our students learn. We can generate testing benefits by prompting student recall with multiple-choice or constructed response questions or by simply asking them to recall information. Further, we can have students record their answers or simply think of them. Finally, we can have students do the recall outside of class, in the middle of class, or at the end of class. The benefits extend beyond the recalled information, improving students' ability to answer inference questions as well as to learn subsequent information. Test-enhanced learning is an approach we often use unconsciously that is worth incorporating into our courses in a more intentional way.

In this section of the book, we have explored four adaptable, generally applicable teaching approaches that can be incorporated into many different course structures to promote our students' learning. In the next section of the book, we turn to four specific pedagogies that can work with and build on these approaches, beginning with lecture.

REFERENCES

Abbott, E.E., 1909. On the analysis of the factors of recall in the learning process. Psychol. Monogr. 11, 159–177.

Bjork, R.A., 1975. Retrieval as a memory modifier: an interpretation of negative recency and related phenomena. In: Solso, R.L. (Ed.), Information Processing and Cognition. Wiley, New York, NY, pp. 123–144.

Butler, A.C., 2010. Repeated testing produces superior transfer of learning relative to repeated studying. J. Exp. Psychol. Learn. Mem. Cogn. 36, 1118–1133.

Butler, A.C., Roediger III, H.L., 2008. Feedback enhances the positive effects and reduces the negative effects of multiple-choice testing. Mem. Cogn. 36, 604–616.

Butler, A.C., Karpicke, J.D., Roediger III, H.L., 2008. Correcting a metacognitive error: feedback increases retention of low-confidence correct responses. J. Exp. Psychol. Learn. Mem. Cogn. 14, 918–928.

Gates, A.I., 1917. Recitation as a factor in memorizing. Arch. Psychol. 6 (40).

James, W., 1890. The Principles of Psychology. Holt, New York, NY.

Kang, S.H.K., McDermott, K.B., Roediger III, H.L., 2007. Test format and corrective feedback modify the effect of testing on long-term retention. Eur. J. Cogn. Psychol. 19, 528–558.

Karpicke, J.D., Blunt, J.R., 2011. Retrieval practice produces more learning than elaborative studying with concept mapping. Science 331, 772–775.

Klionsky, D.J., 2008. The quiz factor. CBE Life Sci. Educ. 7, 265–266.

Larsen, D.P., Butler, A.C., Roediger III, H.L., 2009. Repeated testing improves long-term retention relative to repeated study: a randomized controlled trial. Med. Educ. 43, 1174–1181.

Little, J.L., Bjork, E.L., 2012. The persisting benefits of using multiple-choice tests as learning events. Cogn. Sci. Soc. Retrieved from: <http://mindmodeling.org/cogsci2012/papers/0128/paper0128.pdf> (accessed 11.01.14.).

Lyle, K.B., Crawford, N.A., 2011. Retrieving essential material at the end of lectures improves performance on statistics exams. Teach. Psychol. 38, 94–97.

McDaniel, M.A., Masson, M.E.J., 1985. Altering memory representations through retrieval. J. Exp. Psychol. Learn. Mem. Cogn. 11, 371–385.

McDaniel, M.A., Wildman, K.M., Anderson, J.L., 2012. Using quizzes to enhance summative-assessment performance in a web-based class: an experimental study. J. Appl. Res. Mem. Cogn. 1, 18–26.

Orr, R., Foster, S., 2013. Increasing student success using online quizzing in introductory (majors) biology. CBE Life Sci. Educ. 12, 509–514.

Pazicni, S., Pyburn, D.T. (2014). Intervening on behalf of low-skilled comprehenders in a university general chemistry course. In: Benassi, V.A., Overson, C.E., Hakala, C.M. (Eds.), Applying Science of Learning in Education: Infusing Psychological Science Into the Curriculum. Retrieved from the Society for the Teaching of Psychology web site: <http://teachpsych.org/ebooks/asle2014/index.php>.

Pyc, M.A., Agarwal, P.K., Roediger III, H.L, 2014. Test-enhanced learning. In: Benassi, V.A., Overson, C.E., Hakala, C.M. (Eds.), Applying Science of Learning in Education: Infusing Psychological Science Into the Curriculum. Retrieved from the Society for the Teaching of Psychology web site: <http://teachpsych.org/ebooks/asle2014/index.php>.

Richland, L.E., Kornell, N., Kao, L.S., 2009. The pretesting effect: do unsuccessful retrieval attempts enhance learning? J. Exp. Psychol. Appl. 15 (3), 243–257. Available from: https://doi.org/10.1037/a0016496.

Roediger III, H.L., Butler, A.C., 2011. The critical role of retrieval practice in long-term retention. Trends. Cogn. Sci. 15, 20–27.

Roediger III, H.L., Karpicke, J.D., 2006a. Test-enhanced learning: taking memory tests improves long-term retention. Psychol. Sci. 17, 249–255.

Roediger III, H.L., Karpicke, J.D., 2006b. The power of testing memory: basic research and implications for educational practice. Persp. Psychol. Sci. 1, 181–210.

Roediger III, H.L., Pyc, M.A., 2012. Inexpensive techniques to improve education: applying cognitive psychology to enhance educational practice. J. Appl. Res. Mem. Cogn. 1, 242–248.

Roediger III, H.L., Putnam, A.L., Smith, M.A., 2011. Ten benefits of testing and their applications to educational practice. Psychol. Learn. Motiv. 55, 1–36.

Smith, M.A., Karpicke, J.D., 2014. Retrieval practice with short-answer, multiple-choice, and hybrid tests. Memory 22, 784–802.

Stanger-Hall, K.F., Shockley, F.W., Wilson, R.E., 2011. Teaching students how to study: a workshop on information processing and self-testing helps students learn. CBE Life Sci. Educ. 10, 187–198.

Steele, C.M., 2010. Whistling Vivaldi: How Stereotypes Affect Us and What We Can Do. W.W. Norton & Company, New York, NY.

Wissman, K.T., Rawson, K.A., Pyc, M.A., 2011. The interim test effect: testing prior material can facilitate the learning of new material. Psychon. Bull. Rev. 18, 1140–1147.

Section III

Pedagogy Toolbox

Chapter 8

Lecturing

Lecturing is an essential skill for college science teachers. While we don't want to rely on it as our sole teaching tool, it can serve as a means for instructors to spark students' interest in a new topic, help students develop a framework for a complex subject, and model scientific thinking. Used judiciously and executed well, it can supplement information from readings, frame opportunities for active learning exercises, and help instructors forge relationships with many students at once.

The challenge, of course, is structuring our lectures to maximize our students' learning. This chapter describes principles that can be used as a framework for creating effective lectures and provides examples of specific practices that correspond to these principles and enhance students' learning from lecture.

WHAT ARE PRINCIPLES FOR EFFECTIVE LECTURING?

When considering how to make a lecture effective, it can be useful to revisit a model of memory formation first considered in Chapter 4 (Fig. 8.1). deWinstanley and Bjork (2002) suggest that using this understanding of learning can help us identify practices that are important for effective lectures.

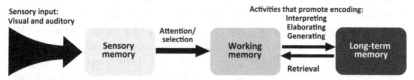

FIGURE 8.1 A model of memory formation can help explain the lecture practices that promote learning. Some sensory input is selected for processing in working memory, which has low capacity and is short-term. Connecting information in working memory to existing knowledge leads to encoding in long-term memory. *The model shown is based on the Atkinson–Shiffrin model, supplemented with information from deWinstanley and Bjork (Atkinson and Shiffrin, 1968; deWinstanley and Bjork, 2002). Although the model can be useful for making teaching choices, it's important to note that it's highly simplified. In particular, short-term working memory and long-term memory function as a continuum rather than as two distinct entities.*

Science Teaching Essentials. DOI: https://doi.org/10.1016/B978-0-12-814702-3.00008-1
109

When listening to a lecture, our students are encountering an ongoing stream of sensory information, some of it related to the topic at hand and some of it completely unrelated. Students select some of this information to pay attention to in working memory, which has a low capacity and is relatively short-term. If they connect that information to their current understanding of a topic, it promotes encoding of the new information into long-term memory. Thus one way to think about what is going on in the learning process is that a network of knowledge—an understanding of what something is and how it works—is being extended. There are several types of processing that lead to memory formation, often called interpretation, elaboration, and generation. Interpretation may be the simplest of these; as the name suggests, it involves students figuring out how the new information fits with their current understanding. Elaboration takes this process a step further; when students elaborate, they are incorporating new information into a broader, coherent narrative and considering its implications. Finally, generation is the process of using partial information to produce new insights, an activity that can be a powerful way to promote formation of long-term memory. New connections may be relatively weak, but the process of retrieving information into working memory can strengthen memories, in part by providing opportunities to make new and reorganized connections.

These observations correspond to three elements that are key for promoting learning from lecture.

- *Attention.* We are constantly bombarded with stimuli, and we necessarily attend to some of these stimuli and ignore others. During a lecture, we want to help our students select key information to process in working memory, allowing them to take the first step toward building knowledge.
- *Interpretation and elaboration.* Interpretation and elaboration are closely related processes that allow information to be incorporated into a student's knowledge structure. Interpretation is the process of fitting new information with what is known, finding associations with preexisting understandings. These preexisting understandings may include academic knowledge, and may also include everyday working knowledge. For example, a student learning about the strength of hydrogen bonding may relate it to preexisting academic knowledge about covalent bonds, but may also relate it to the pain of previous belly flops in the pool. Elaboration takes the formation of these connections a step further, incorporating new information into a broader, coherent narrative. It involves the formation of connections across course content as well as consideration of implications. For example, when a student is learning about the citric acid cycle, she might first consider how the enzyme isocitrate dehydrogenase fits within the cycle, interpreting its role in converting isocitrate to α-ketoglutarate. As she elaborates on that understanding, however, she may also recognize that the reaction proceeds with a highly

negative ΔG and is thus a likely site of regulation; that it oxidizes a carbonyl, and thus requires the cofactor NAD +; that it is located in the mitochondrial matrix and thus directly "feeds" the electron transport chain. By thinking about the information in different ways, the student is making interconnections that form a richer network, enhancing her ability to apply the information in different contexts and increasing the number of retrieval routes by which she can access the information.

- *Retrieval practice.* Recalling information is one of the most powerful ways we know to enhance learning. Interpretation and elaboration allow us to incorporate new information into existing knowledge networks, but retrieval practice enhances the strength of the connections in the networks and can help forge new, previously unrecognized connections. Thus, it not only helps us remember knowledge more quickly and reliably, it also extends our understanding of how that knowledge fits within our larger mental models. Retrieval practice can take a number of forms, from free recall of a class session, to explaining a process to a colleague, to using flashcards, and its use—intentional or unintentional—plays a role in the development of every memory.

These elements can provide a powerful framework for considering how to adapt and supplement our lectures to maximize students' learning. As we consider ways to maximize and direct students' attention, promote interpretation and elaboration, and foster retrieval practice, it's important to note that the elements are not entirely separable. For example, using a case study to contextualize a particular concept may increase student interest and attention, but also provides opportunities for connections between the concept and real-life scenarios. Likewise, asking students to apply recently acquired knowledge to a new scenario promotes both retrieval practice and elaborative processing. Thus many good lecture practices target memory formation at multiple levels.

WHAT ARE PRACTICES TO PROMOTE ATTENTION?

The key to thinking about attention is to consider two elements: the steps you'll take to encourage your students to pay attention to the lecture instead of some other stimulus (e.g., worry about another deadline, waves of sleepiness), and the steps you'll take to help them select the right information from the lecture.

1. *Use good public speaking skills.* It may seem fairly obvious, but a lecture is an example of public speaking, and the approaches that make a public talk engaging will help ensure that your students pay attention to what you have to say.
   - *Make eye contact with the students.* As social animals, humans tend to respond to cues that make them feel closer to a speaker (also

known as immediacy cues). Using signals like eye contact that demonstrate to students that they belong in the classroom and that you're speaking to them increases their attention and thus their learning, particularly for students who have not yet developed strong self-regulation skills (Bolkan et al., 2017). It's a simple step that can have large impact.

• *Think of your lecture as a narrative.* As we move through a lecture, we are often telling a story: the story of DNA replication, electrophilic attack, or the effect of a magnetic field. In some ways, it's a nontraditional story—the players are often not animals, the actions nonautonomous—but it can have the characteristics of a plot, with initial actions that set up later events, tension, climax, and resolution. Humans have a long history as storytellers, and we tend to pay attention to and remember stories. Thinking about a lecture as a story can help you take advantage of this tendency to capture student attention and potentially enhance learning (Glonek and King, 2014; Bryant and Harris, 2011).

• *Use movements, facial expressions, pauses, changes in pitch, and injections of enthusiasm to emphasize key points.* Good storytellers know that changes in delivery help capture and maintain listeners' interest—and all of us know that monotone lectures rapidly lead us to think about our grocery list. These types of nonverbal communication can have significant impact on students' learning (York, 2013). Using hand gestures, movements of the head, or exclamations can emphasize critical points and help point student attention to the most important elements of a lecture. Pauses also have the potential to refocus attention and can be particularly powerful; if students have been listening to a flow of words, then a deliberate pause can punctuate that flow to help students recognize that something important is at hand. Likewise, facial expressions—widened eyes, lifted eyebrows—can call attention to unexpected, unintuitive observations that have proven to be particularly important. Perhaps surprisingly, selective injection of instructor enthusiasm can also act as an attention-focusing device. Wood observed that uniform instructor enthusiasm did not increase student learning from lecture, but that enthusiasm about specific elements increased student attention and memory of those elements (Wood, 1999). Finally, large movements that serve to illustrate an aspect of the lecture (such as how a DNA polymerase can proofread) can have particular value for focusing attention in lectures that are describing invisible physical phenomena.

2. *Use one or more mechanisms to help students identify the key elements to take away from the lecture.* One of the most useful recognitions we can make in our teaching has to do with differences in the ways we and our students encounter information in our field. Even when we are

teaching something outside our comfort zone, we have a rich understanding of our field as a whole. We know how the course we are teaching fits into the larger discipline, and we know the norms of the discipline. If we're lucky enough to be teaching a class we've taught several times before or that is closer to our area of expertise, we have an even more robust understanding. Our students, on the other hand, have a relatively shallow understanding of what our field is, how it builds knowledge, and what the norms are for conveying information. Well-structured lectures that intentionally focus student attention can help bridge this gap, helping students recognize the important knowledge and abilities they should be constructing from the lecture. The following practices can help provide this structure.

- *Have students do preclass work that activates prior knowledge and leaves them with questions.* Schwartz and Bransford (1998) published a now-classic study on the impact of different types of preclass work on learning from lecture. They compared the effects of analyzing contrasting cases to summarizing a text about the relevant concepts prior to lecture in a cognitive psychology class. For the analysis condition, students were prompted to find and graph revealing patterns for the case; in the summarization condition, they wrote 2–3 page summaries of the readings. Students who performed the analysis and then heard a lecture helping them to understand the underlying concepts were able to make more than twice as many predictions from a related case study. Further, students who heard no lecture but instead analyzed additional case studies performed about as well as students who summarized and heard a lecture. The results suggest that preclass work in which students look for patterns and ask questions creates a "time for telling" that enhances their learning from a subsequent lecture.

- *Share learning objectives with students at the beginning of a lecture.* One way to focus student attention on the skills and knowledge you want them to develop during lecture is to highlight those elements at the beginning of class. Learning objectives are statements about what you want students to be able to do at the end of a learning session (be that a single class period, a unit, or a whole semester). They have a student-focused verb as well as a description of the subject matter. For example:

- Be able to sketch a force diagram for suspended objects.
- Be able to classify inheritance problems according to whether they demonstrate complete dominance, incomplete dominance, or codominance.
- Be able to describe and explain the electron movements in Michael addition reactions.

These statements give students a tool to help them focus on the key elements they should take from lecture, directing their attention where the instructor wants it to be. They also give the instructor an opportunity to highlight that examples from the lecture are often just that—examples—and that students should also consider how the concept applies in other examples.

- *Give students an ungraded pretest before lecture.* Multiple choice pretests prior to lecture can help focus student attention during lecture, enhancing learning of both the pretested information and related information. Little and Bjork (2016) investigated the impact of multiple choice pretesting in a lab setting. They compared the effect of a multiple choice pretest to a cued recall pretest (which used the same questions without alternatives) and a prefact study control, in which students read statements that answered the pretest questions. They found that the multiple-choice pretestenhanced learning, improving student performance on the posttest by about 35% in comparison to the other conditions. Notably, the posttest examined student learning of information related to the pretest but did not replicate the pretest.

3. *Consider the technology you use.* Many times, we tend to adopt the same technology that we use for a research talk for our classrooms: powerpoint or some other presentation software with prepared slides that we refer to throughout the lecture. This approach offers the advantages of allowing the instructor to incorporate critical images as well as video and "clicker" questions. It can, however, encourage instructors to move too quickly and foster a more passive experience for students. Presenting lecture slides via tablet PCs with digital inking allows the incorporation of graphical images but also creates a more "real-time" experience, with instructors adding notations to the slides to extend and emphasize existing information. The notations can be recorded and shared with students after class. Some work suggests that this type of delivery improves students' perceptions of lecture interest, pace, and understanding (Choate et al., 2014; Revell, 2014).

WHAT ARE PRACTICES TO HELP STUDENTS MAKE CONNECTIONS?

Since encoding information into long-term memory requires the formation of connections with existing knowledge, incorporating cues for students to interpret and elaborate on lecture information is key for learning from lecture. The following practices can give students tools and opportunities to build new information from lecture into their understanding.

1. *Activate prior knowledge.* Helping students draw relevant knowledge back into their working memory before providing new information can

increase learning (Bower et al., 1969). The process can be simple. Projecting an image and asking students relevant questions can be powerful. For example, the instructor could project a map of catabolic pathways and ask students to take 90 seconds to note what they remember about the pathways, or could project a line drawing of a particular molecule and ask what kinds of bonds it can form. Alternatively, giving a low- or no-stakes quiz that asks students to recall relevant information makes that information more accessible for formation of new connections. Another option that builds into the practice of making the lecture organization explicit involves the development of a concept map. Concept maps (described more fully in Chapter 4) are visual maps that connect key elements within a topic through arrows that describe relationships; an example is shown in Fig. 8.2. By taking a few minutes at the beginning of a lecture to guide students in constructing a collective concept map about a topic, you can give your students a starting point to begin extending their knowledge.

2. *Make the organization of your lecture explicit.* As we prepare our lectures, we think carefully about organization. What are the overarching principles? What are relevant examples? What ideas flow from others, and why? Making that organization visible to students by providing an outline or map can help them make the connections that increase their learning. The concept map described above as a means to activate prior knowledge can be extended during class, providing a visual tool that can help students interpret new information to fit within an existing framework. That map can be generated by the class and thus have the advantage of also activating prior knowledge, as described above, but it can also have value if provided by the instructor. For example, Cutrer and colleagues found that using an expert concept map to illustrate the organization of a lecture on respiratory failure for resident physicians improved learning even within this highly motivated set of adult learners

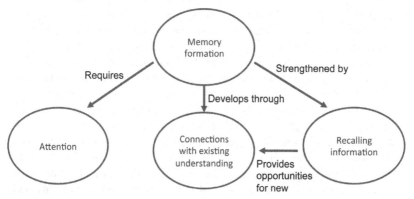

FIGURE 8.2   Example concept map.

(Cutrer et al., 2011). Alternatively, providing an outline of the lecture on the board or in posted documents can also help students see important connections. This approach can be particularly helpful if the outline is shared with students for notetaking purposes. Specifically, providing students with outlines for notetaking has been shown to improve student notetaking, memory, and test performance, particularly if the provided outline does not provide the full detail of the lecture (Kiewra et al., 1995; Morgan et al., 1988).

3. *Use visual images that convey meaning and connections.* Visual images are powerful aids to memory that can capture multiple ideas and convey connections among them. Not only does this give students a tool to make connections, the image can also give students a tool for "chunking" multiple ideas into a meaningful whole. Images that are specifically selected to show connections or multiple relationships (e.g., graphs or diagrams) are particularly helpful in this regard; iconic or entertaining images like photos may capture student interest but don't necessarily provide a way to highlight connections. For example, an image that shows a cell signaling cascade condenses a large number of reactions to a single chunk and highlights the inherent relationships within the cascade. It is important to guide students through the images during the lecture for two reasons. First, students are often unfamiliar with the conventions we use for figures within our disciplines and do not readily interpret them. Second, images that have a large amount of information about meaning and connections can be overwhelming, generating a cognitive load that overwhelms the working memory. By guiding students through the image during the lecture, the instructor can help focus on the important elements conveyed in the image to help students build it into a coherent whole.

4. *Incorporate pauses that give students space to make connections.* Lectures are often seen as a pedagogy that passes words over students without their participation (see, e.g., Mazur, 2009). In articles like these, lecture is typically defined as "continuous exposition by the teacher" (Freeman et al., 2014), an approach that virtually ensures that students zone in and out and don't have time to do much processing in class. However, lectures that are structured to help students make connections and that are punctuated with opportunities for them to do so can be effective and relatively efficient ways to help students learn sophisticated ideas (Schwartz and Bransford, 1998). There are many ways to pause during lecture to provide opportunities for students to make connections; three simple ways are the pause procedure, think—pair—share, and minute papers. In the pause procedure, the instructor pauses for 2 minutes every 12—18 minutes, encouraging students to discuss and rework notes in pairs, after which they are invited to ask the instructor questions. This approach has been shown to significantly increase learning when

compared to lectures without the pauses (Rowe, 1980; Ruhl et al., 1987). In the think—pair—share method, the instructor poses a question, gives students a brief period to consider it individually, instructs them to talk to their neighbors about their reasoning, and then solicits and discusses answers within the whole class. This classic active learning technique gives students the chance to articulate their interpretations and to supplement them with ideas from their colleagues, providing opportunities for processing that lead to learning. Minute papers are brief writing exercises in which students respond to an instructor's prompt; instructors can ask general questions such as, "What were the three main ideas in today's lecture?" or "What is the most confusing idea from this section of the lecture?" or can ask more specific questions about the lecture content. Like the think—pair—share approach, this approach encourages students to articulate and examine newly formed connections (Angelo and Cross, 1993; Handelsman et al., 2007).

These approaches may prove most effective if you take some time at the beginning of the semester to talk to your students about how learning works. Helping them understand the benefits of making connections between new information and their previous understanding may help increase their willingness to engage with active learning interludes and other additions to traditional lecture (Stanger-Hall et al., 2011; Seidel and Tanner, 2013).

WHAT ARE PRACTICES TO STRENGTHEN AND EXTEND MEMORY?

Retrieval practice is one of the most effective tools that students can use to promote their learning. In recalling information, they strengthen memory and provide opportunities to form new connections in another round of processing. Telling your students about this effect, perhaps sharing data with them (see Chapter 7 for useful graphs), can give them a powerful tool to enhance their learning. In addition, however, you may consider opportunities to promote retrieval practice in your lecture.

1. *Use the last 5 minutes of class for students to recall lecture information.* Lyle and Crawford examined the effects of retrieval practice on student learning in undergraduate statistics class (Lyle and Crawford, 2011). In one section of the course, students were instructed to spend the final 5—10 minutes of each class period writing answers, from memory, to two to four questions that required them to retrieve information about the day's lecture. The students in this section of the course performed about 8% higher on exams over the course of the semester than students in sections that did not use the retrieval practice method, a statistically significant difference.

2. *Use the first 2—5 minutes of class for students to recall information from a previous class.* At the beginning of class, instructors can use a quiz or

free writing exercise to encourage students to recall relevant information from the previous class. If, is as often the case, the information acts as a starting point for the current lecture, then this approach also has the benefit of activating relevant prior knowledge.

Lectures are primarily about students encountering, processing, and organizing new information; retrieval practice is part of the work that they will do outside of class to help solidify and extend their understanding. Because retrieval practice is so powerful and can merge with techniques that promote learning in other ways (e.g., activating prior knowledge, elaborating on connections with existing knowledge), however, it can be a helpful addendum to a well-structured lecture.

CONCLUSION

A model of memory formation can provide a useful framework for considering how to structure and supplement our lectures to enhance students' learning. When we consider ways to maximize and direct students' attention, foster the formation of connections between new information and existing knowledge, and promote retrieval practice during our lectures, we're doing more than "continuous exposition"; we're using structures and approaches that help our students develop understanding of the complex subjects we teach. Thus, the punctuated lectures that we develop using this framework can be a useful tool in our teaching toolkit.

In this section of the book, we are exploring four specific pedagogies that incorporate and build on the fundamental teaching and learning approaches of active learning, group work, metacognitive practices, and retrieval practice. In this chapter, we've considered lecturing, the pedagogy most familiar to most of us; in the next, we turn to the flipped classroom.

REFERENCES

Angelo, T.A., Cross, K.P., 1993. Classroom Assessment Techniques: A Handbook for College Teachers. Jossey-Bass, San Francisco, CA.
Atkinson, R.C., Shiffrin, R.M., 1968. Human memory: a proposed system and its control processes. In: Spence, K.W., Spence, J.T. (Eds.), The Psychology of Learning and Motivation, vol. 2. Academic Press, New York, NY, pp. 89–195.
Bolkan, S., Goodboy, A.K., Myers, S.A., 2017. Conditional processes of effective instructor communication and increases in students' cognitive learning. Commun. Educ. 66, 129–147.
Bower, G.H., Clark, M.C., Lesgold, A.M., Winzenz, D., 1969. Hierarchical retrieval schemes in recall of categorized word lists. J. Verbal Learn. Verbal Behav. 8, 323–343.
Bryant, L., Harris, R., 2011. Using storytelling to increase interest and recollection in finance concepts. J. Instr. Pedagogies 6, 1–11.
Choate, J., Kotsanas, G., Dawson, P., 2014. Exploring tablet PC lectures: lecturer experiences and student perceptions in biomedicine. Aust. J. Educ. Technol. 30 (2), 167–183.

Cutrer, W.B., Castro, D., Roy, K.M., Turner, T.L., 2011. Use of an expert concept map as an advanced organizer to improve understanding of respiratory failure. Med. Teach. 33, 1018–1026.

deWinstanley, P.A., Bjork, R.A., 2002. Successful lecturing: presenting information in ways that engage effective processing. New Dir. Teach. Learn. 89, 19–31.

Freeman, S., Eddy, S.L., McDonough, M., Smith, M.K., Okoroafor, N., Jordt, H., et al., 2014. Active learning increases student performance in science, engineering, and mathematics. Proc. Nat. Acad. Sci. U.S.A. 111, 8410–8415.

Glonek, K.L., King, P.E., 2014. Listening to narratives: an experimental examination of storytelling in the classroom. Int. J. Listen. 28, 32–46.

Handelsman, J., Miller, S., Pfund, C., 2007. Scientific Teaching. W.H. Freeman, New York, NY.

Kiewra, K.A., Benton, S.L., Kim, S.-I., Risch, N., Christensen, M., 1995. Effects of note-taking format and study technique on recall and relational performance. Contemp. Educ. Psychol. 20, 172–187.

Little, J.L., Bjork, E.L., 2016. Multiple-choice pretesting potentiates learning of related information. Mem. Cognit. 44, 185–1101.

Lyle, K.B., Crawford, N.A., 2011. Retrieving essential material at the end of lectures improves performance on statistics exams. Teach. Psychol. 38, 94–97.

Mazur, E., 2009. Farewell, lecture? Science 323, 50–51.

Morgan, C.H., Lilley, J.D., Boreham, N.C., 1988. Learning from lectures: the effect of varying the detail in lecture handouts on note-taking and recall. Appl. Cogn. Psychol. 2, 115–122.

Revell, K.D., 2014. A comparison of the usage of tablet PC, lecture capture, and online homework in an introductory chemistry course. J. Chem. Educ. 91, 48–51.

Rowe, M.B., 1980. Pausing principles and their effects on reasoning in science. New Dir. Commun. Coll. 31, 27–34.

Ruhl, K., Hughes, C.A., Schloss, P.J., 1987. Using the pause procedure to enhance lecture recall. Teach. Educ. Spec. Educ. 10, 14–18.

Schwartz, D., Bransford, J., 1998. A time for telling. Cogn. Instr. 16 (4), 475–522.

Seidel, S.B., Tanner, K.D., 2013. "What if students revolt?"—considering student resistance: origins, options, and opportunities for investigation. CBE Life Sci. Educ. 2, 586–595.

Stanger-Hall, K.F., Shockley, F.W., Wilson, R.E., 2011. Teaching students how to study: a workshop on information processing and self-testing helps students learn. CBE Life Sci. Educ. 10, 187–198.

Wood, A.M., 1999. The Effects of Teacher Enthusiasm on Student Motivation, Selective Attention, and Text Memory (Unpublished doctoral dissertation). University of Western Ontario.

York, D., 2013. Investigating a Relationship Between Nonverbal Communication and Student Learning (Doctoral dissertation). Lindenwood University.

Chapter 9

Flipping the Classroom

The flipped classroom has become an important educational phenomenon, driven in part by high profile publications in *The New York Times* (Fitzpatrick, 2012), *The Chronicle of Higher Education* (Berrett, 2012), and *The Atlantic* (Bogost, 2013), among others. In essence, flipping the classroom means that students gain first exposure to new material outside of class, often (but not always) through videos, and then use class time to do the harder work of assimilating that knowledge, perhaps through problem-solving, discussion, or debates. In terms of Bloom's revised taxonomy (Fig. 9.1; Anderson and Krathwohl, 2001), this means that students are doing the lower levels of cognitive work (gaining knowledge and comprehension) outside of class, and focusing on the higher forms of cognitive work (application, analysis, synthesis, and/or evaluation) in class, where they have the support of their peers and instructor. This model contrasts from the traditional model in which "first exposure" occurs via lecture in class, with students assimilating knowledge through homework, and it's this reversal that produced the term flipped classroom.

As an educational phenomenon, the flipped classroom has staying power because it relies on two important tools of modern teaching: it can leverage the potential of interactive virtual tools outside of class as well as the known value of active learning approaches in class. The combination has the potential to create powerful learning experiences.

This chapter describes the development of the flipped classroom idea, the role it can play in promoting learning, and some of the evidence that it

FIGURE 9.1 Bloom's revised taxonomy of cognitive processes (Anderson and Krathwohl, 2001).

Science Teaching Essentials. DOI: https://doi.org/10.1016/B978-0-12-814702-3.00009-3

improves student performance. The chapter ends with the key elements of the flipped classroom and accompanying practical advice that can help make the flipping experience productive for both faculty and students.

HOW WAS THE FLIPPED CLASSROOM IDEA DEVELOPED?

The flipped classroom approach has been used for years in some disciplines, notably within the humanities. Barbara Walvoord and Virginia Johnson Anderson promoted the use of this approach in their book Effective Grading (1998). They proposed a model in which students gain first exposure learning prior to class and focus on the processing part of learning (synthesizing, analyzing, problem-solving, etc.) in class. To ensure that students do the preparation necessary for productive class time, Walvoord and Anderson proposed an assignment-based model in which students produce work (writing, problems, etc.) prior to class. The students receive productive feedback through the processing activities that occur during class, reducing the need for the instructor to provide extensive written feedback on the students' work. Walvoord and Anderson describe examples of how this approach has been implemented in history, physics, and biology classes, suggesting its broad applicability.

Maureen Lage, Glenn Platt, and Michael Treglia described a similar approach as the inverted classroom, and reported its application in an introductory economics course in 2000 (Lage et al., 2000). Lage, Platt, and Treglia designed an inverted classroom in which they provided students with a variety of tools to gain first exposure to material outside of class: textbook readings, lecture videos, PowerPoint presentations with voice-over, and printable PowerPoint slides. To help ensure student preparation for class, students were expected to complete worksheets that were periodically but randomly collected and graded. Class time was then spent on activities that encouraged students to process and apply economics principles, ranging from mini-lectures in response to student questions to economic experiments to small group discussions of application problems. Both student and instructor response to the approach was positive, with instructors noting that students appeared more motivated than when the course was taught in a traditional format.

Catherine Crouch and Eric Mazur described a modified form of the flipped classroom, which they term peer instruction, in 2001. Like the other approaches described, the peer instruction model requires that students gain first exposure prior to class, and uses assignments (in this case, quizzes) to help ensure that students come to class prepared. Class time is structured around alternating mini-lectures and conceptual questions. Importantly, the conceptual questions are not posed informally and answered by student volunteers as in traditional lectures; instead, all students must answer the conceptual question, often via "clickers," or handheld personal response systems, that allow students to answer anonymously and that allow the instructor to see (and display) the class data immediately. If a large fraction of the class (usually between 30% and 65%) answers incorrectly, then students reconsider the question in small groups while instructors circulate to promote productive discussions. After discussion, students answer the conceptual question again. The instructor provides feedback, explaining the correct answer and following up with related questions if appropriate. The cycle is then repeated with another topic, with each cycle typically taking 13−15 minutes.

More recently, the flipped classroom has become associated with preparation via online learning materials (Strayer, 2012). For example, the flipped classroom described by McLean and colleagues used online learning modules that included videos, formative quizzes, and other interactive elements (McLean et al., 2016). By leveraging the power of their webcam, autograding quizzes with targeted feedback, blogs, and other interactive tools, instructors can personalize the preclass component of the flipped classroom. Coupled with the demonstrated potential of active learning approaches in the classroom, this combination provides the opportunity for instructors to design highly effective learning experiences for their students.

WHAT'S THE THEORETICAL BASIS? OR, WHY SHOULD IT WORK?

The potential of the flipped classroom to promote learning can be explained by three educational theories. Two of these theories relate to the active learning approaches used during the in-class portion of the flipped classroom, while the third relates to preclass preparation.

Constructivist learning theory emphasizes that individuals learn through building their own knowledge, connecting new ideas and experiences to existing knowledge to form new or enhanced understanding (Bransford et al., 2000). The theory, developed by Piaget and others, posits that learners can either assimilate new information into an existing framework, or can modify that framework to accommodate new information that contradicts prior understanding. Active learning approaches, a key element of the flipped classroom, are based in constructivist learning theory. Approaches that promote active learning often explicitly ask students to make connections between new information and their current mental models, extending their understanding. In other cases, teachers may design learning activities that allow students to confront misconceptions, helping students reconstruct their mental models based on more accurate understanding. In either case, active learning promotes the kind of cognitive work identified as necessary for learning by constructivist learning theory.

Active learning approaches also often embrace the use of cooperative learning groups, a constructivist-based practice that places particular emphasis on the contribution that social interaction can provide. Lev Vygotsky's work elucidated the relationship between cognitive processes and social activities and led to the sociocultural theory of development, which suggests that learning takes place when students solve problems beyond their current developmental level with the support of their instructor or their peers (Vygotsky, 1978). Thus, active learning approaches that rely on group work rest on this sociocultural branch of constructivist learning theory, leveraging peer−peer interaction to promote students' development of extended and accurate mental models.

Abeysekera and Dawson (2015) propose that the preclass preparation associated with the flipped classroom also provides an opportunity for students to manage cognitive load. According to cognitive load theory, working memory is limited and subject to the intrinsic load of the subject, the germane load of learning activities that help promote understanding, and extraneous load (Clark et al., 2005). By encouraging students to regulate their pace and self-monitor learning during preclass work, instructors may be able to help their students better manage their cognitive load and therefore increase learning.

IS THERE EVIDENCE THAT IT PROMOTES LEARNING?

In essence, the flipped classroom is the use of preclass preparation and in-class active learning. Thus, the voluminous literature on active learning approaches is informative (see Box 9.1). A randomized controlled trial in a junior level biochemistry course emphasizes the relevance of these data: students were randomly distributed to sections of the course that differed in class activities (lecture + instructor demonstration of problem-solving vs lecture + active student problem-solving) and preclass preparation (reading vs video) (Stockwell et al., 2015). Students preferred preparation using videos but showed no difference in exam performance in the two preparation methods. The active learning approach, however, significantly improved student learning. Jensen and colleagues also found that active learning was the key element of flipped classroom pedagogy (Jensen et al., 2015). They varied the order of instructor involvement—during content attainment or content application—but used active learning approaches in both conditions. They found no significant difference in students' performance in the two classes, suggesting that reported benefits from flipped classes derive from the use of active learning.

There are also direct comparisons of flipped and traditional classrooms in several disciplines, examining outcomes that range from exam performance to study strategies used.

- Lockman and colleagues compared student performance on an objective structured clinical exam for PharmD students learning pain therapeutics

BOX 9.1 Active Learning: Powerful Evidence That it Works

In 1998, Richard Hake gathered data on 2084 students in 14 introductory physics courses taught by traditional methods (defined by the instructor as relying primarily on passive student lectures and algorithmic problem exams), allowing
him to define an average gain for students in such courses using pre-/posttest
data (Hake, 1998). Hake then compared these results to those seen with interactive engagement methods, defined as "heads-on (always) and hands-on (usually)
activities that yield immediate feedback through discussion with peers and/or
instructors" (p. 65) for 4458 students in 48 courses. He found that students taught
with interactive engagement methods exhibited learning gains almost two standard deviations higher than those observed in the traditional courses
(0.48 ± 0.14 vs 0.23 ± 0.04).

Carl Wieman and colleagues have also published evidence that using active
learning during class can produce significant learning gains (Deslauriers et al.,
2011). Wieman and colleagues compared two sections of a large-enrollment
physics class. The classes were both taught via interactive lecture methods for
the majority of the semester and showed no significant differences prior to the
experiment. During the 12th week of the semester, one section was "flipped,"
with first exposure to new material occurring prior to class via reading assignments and quizzes, and class time devoted to small group discussion of clicker
questions and questions that required written responses. Although class discussion was supported by targeted instructor feedback, no formal lecture was
included in the experimental group. The control section was encouraged to read
the same assignments prior to class and answered most of the same clicker questions for summative assessment but were not intentionally engaged in active
learning exercises during class. During the experiment, student engagement
increased in the experimental section (from $45 \pm 5\%$ to $85 \pm 5\%$ as assessed by
four trained observers) but did not change in the control section. At the end of
the experimental week, students completed a multiple choice test, resulting in
an average score of $41 \pm 1\%$ in the control classroom and $74 \pm 1\%$ in the
"flipped" classroom, with an effect size of 2.5 standard deviations. Although the
authors did not address retention of the gains over time, this dramatic increase in
student learning supports the use of the flipped classroom model.

through a traditional or a flipped format, finding that the students in the
flipped classroom performed 12% higher on the exam, with particular
improvement on segments of the exam assessing higher cognitive levels
(Lockman et al., 2017). The students in the flipped class also had a 5%
higher average on the pain therapeutics content of the end-of-semester
multiple choice exam, while other sections of the exam (taught with traditional methods in both sections) did not differ between the two sections.

- In a five-year study, Gross and colleagues found that students in a physical chemistry course exhibited significantly better exam performance in a
flipped class than in a traditional class in spite of spending $33-50\%$ less

time in class, a difference due in part to more consistent and accurate engagement with online homework (Gross et al., 2015).

- Van Vliet and colleagues found that adopting a flipped classroom approach in a neuroscience class increased students' critical thinking and use of peers as learning resources (Van Vliet et al., 2015). Students in the study took two consecutive courses, the first taught with traditional lecture pedagogy and the second taught with 20% of the class meetings "flipped." Students completed the Motivated Strategies for Learning Questionnaire at the beginning and end of each of the two courses, exhibiting no changes in most components but increases in the critical thinking and peer learning components after engaging in the flipped course.

- McLean et al. found that students reported less multitasking both during content acquisition and content application when comparing a flipped class to their favorite lecture class (McLean et al., 2016). They also reported using a variety of active learning strategies when completing the online learning modules, ranging from writing notes (49% of respondents) to identifying key concepts (16%) to rereading (24%) and taking an optional quiz (25%). 70% of the students agreed or strongly agreed that the class helped improve their independent learning skills.

These results are promising, but it's worth asking whether different student populations respond differently to flipped classroom pedagogy. Eddy and Hogan found that a flipped classroom approach improved exam performance for all students but provided particular benefits for black and first-generation college students (Eddy and Hogan, 2014). They compared a traditional lecture approach to a "moderate structure" approach that involved preclass reading with ungraded guiding questions, a graded preclass quiz, and in class active learning approaches for an introductory biology course, finding that the flipped approach increased exam performance of all students by 3.2%, black students by 6.3%, and first-generation students by 6.1%. Student survey responses also indicated that students spent more time preparing for the flipped class and that they perceived a stronger sense of classroom community, both of which can benefit all students. Thus, while it's not a panacea, the evidence suggests that the use of the flipped classroom, when done well, can help our students learn.

TAKING PRACTICAL STEPS: WHAT ARE THE KEY ELEMENTS OF THE FLIPPED CLASSROOM?

1. *Before class:* Provide students opportunity and incentive to prepare in a meaningful and productive way.
   - *The basics: provide an opportunity for students to gain first exposure prior to class.* The mechanism used for first exposure can vary, from textbook or article readings, to lecture videos, to podcasts or

screencasts. Students often appreciate well-chosen digital materials that can provide visuals and explanations in a way that are hard to replicate in text, and these materials can provide benefits for some topics (see chapter on Effective Educational Videos), but most of the examples of flipped classrooms described in this chapter use preclass reading for student preparation (see Stockwell et al. (2015) for a direct comparison of reading and video as a preparation method).

- *Provide a mechanism to help students process the first exposure.* Our students don't always get what we want them to out of preclass work, sometimes because they don't yet understand how to do the type of reading and watching needed for our field, and other times because they are doing the work too passively. By providing guiding questions or diagrams or other graphic organizers for notetaking, we can help students do the preclass work in a way that helps them come to class better prepared to engage. For example, the moderate structure class described by Eddy and Hogan provided students with ungraded guided reading questions that prompted students to consider the meaning of particular figures and to illustrate important ideas from the text (Eddy and Hogan, 2014). Examples of assignments are shown in Table 9.1.

- *Provide an incentive for students to prepare.* Just as it's important to give students tools to help them get the most out of preclass preparation, it's also important to provide incentives to make sure they *do* the preparation. Like all of us, students prioritize their activities, and providing an incentive for them to do the preparation conveys that we consider that work a priority. The form of the incentive can vary; some instructors have students submit discussion questions online before class, others use auto-graded quizzes, while others use completed graphic organizers as "entry tickets" for class. In many cases, grading for completion rather than accuracy can be sufficient,

---

TABLE 9.1 Examples of Preclass Work to Help Students Process

Complete a 3−2−1 on the assigned video: What were the three most important ideas? What are two things about which you're unclear? What is one question you'd like to ask the author to take it further? Post your answers to the class blog the evening before class.

Based on the reading, draw a concept map that relates the following concepts: electric field, electric force, electric charge, and Gauss' law. Bring your map to class to discuss with colleagues.

Sketch the citric acid cycle. For each step, write a short description of the chemical modification that is occurring. Circle any carbon that is oxidized in a particular step, and list the cofactor that is concomitantly reduced.

particularly if class activities will provide students with the kind of feedback that grading for accuracy usually provides.

- *Provide a means to assess student understanding.* The preclass assignments that students complete as evidence of their preparation can also help both the instructor and the student assess understanding. Preclass online quizzes can allow the instructor to practice Just-in-Time Teaching (Novak et al., 1999), which basically means that the instructor tailors class activities to focus on the elements with which students are struggling. If automatically graded, the quizzes can also help students pinpoint areas where they need help. Preclass worksheets can also help focus student attention on areas with which they're struggling, and can be a departure point for class activities, while preclass writing assignments help students clarify their thinking about a subject, thereby producing richer in-class discussions. Importantly, much of the feedback students need is provided in class, reducing the need for instructors to provide extensive commentary outside of class (Walvoord and Anderson, 1998). In addition, many of the activities used during class time (e.g., clicker questions or debates) can serve as informal checks of student understanding.

A single assignment can, of course, serve all of these functions, helping students process their first exposure, incentivizing them to do the first exposure, and serving to help students and the instructor evaluate their understanding. The important thing is to keep in mind that each function—processing, incentive, and assessment—is important to consider when constructing your preclass work.

2. *During class:* Structure activities that focus on higher level cognitive functions and help students leverage their peers and instructor to reach the learning goals.

   If the students gained basic knowledge outside of class, then they need to spend class time to promote deeper learning. The activity will depend on the learning goals of the class and the culture of the discipline. For example, Lage, Platt, and Treglia described experiments students did in class to illustrate economic principles (Lage et al., 2000), while Mazur and colleagues focused on student discussion of conceptual "clicker" questions and quantitative problems focused on physical principles (Crouch and Mazur, 2001). In other contexts, students may spend time in class engaged in debates, data analysis, or synthesis activities. The key is that students are using class time to deepen their understanding and increase their skills at using their new knowledge.

   - *The basics: articulate your learning objectives to help focus your in-class work.* It's critical to design your instruction around learning goals, both at the level of the class as a whole and for specific units and even class sessions. By stating your goals for your students in terms of what they should be able to do with the content you are covering, you can help clarify the skills and understanding you want them to develop while working in class. You may wish to have learning objectives for both preclass work and in-class work. For example, a session on the citric acid cycle might have the following learning goals for preclass and in-class work:

- After doing the reading, you should be able to describe the basic eight steps of the citric acid cycle, including the enzymes that catalyze the reactions.
- After class, you should be able to analyze the impact of a poison and develop a model of its impact on the citric acid cycle function.

Chapter 2 describes the use of learning goals to shape whole course design, and there is a Spotlight following Chapter 2 that provides guidance on writing learning objectives to structure specific class units and sessions.

- *Design in-class activities that allow students to work with their colleagues on challenging problems.* The nature of the problems students tackle will depend on the course content, learning objectives, student population, and physical location. A genetics class of 30 upper-level students focusing on explaining mitochondrial diseases will do very different work than a general chemistry class of 200 first-year students developing their understanding of acid–base equilibria. For example, the genetics students may work in persistent small groups over several days to create patient education materials describing mitochondrial transmission, probabilities of disease development, and treatment options, while the chemistry students may use clicker questions to shape short pair- and trio-discussions throughout a class period. In each case, however, students should be given an opportunity to work with each other on problems that are relevant for their learning and are challenging for them given their level of development and the course goals. Chapter 5 on group work describes several group structures that can be used in flipped classrooms.
- *Provide enough guidance that students can focus on intellectual rather than procedural challenges.* We have all been in classes or other situations where the instructor says to get started on the work, and the students turn to each other and wonder what they're supposed to do. Class time is precious, and we want our students to spend it doing the intellectual work that will advance their understanding. This may require that you spell out the steps that you want the students to take more explicitly than feels needed. A good test of whether your instructions are clear enough is to ask an undergraduate or graduate student to read them and indicate if they would know how to get started.

- *Include opportunities for instructor guidance and feedback.* Student—student interaction is very helpful for promoting student learning, but the instructor has a key role to play in any college classroom. Students need to know if their thinking is moving in the right direction, and they benefit from hearing instructors model the type of thinking that characterizes the discipline. Plan for students to "report out" on their progress to give you a chance to redirect as needed and to give the whole class an opportunity to hear your feedback.
- *Plan your timing.* Active learning approaches can expand to fill more time than expected. The best defense against this expansion is to plan the timing of a set of activities and, generally, to stick to it. For example, you might allot 4 minutes for a think—pair—share: 30 seconds for reading the question; 30 seconds for thinking individually; 60 seconds for partner discussion; 90 seconds for sharing out; 30 seconds for instructor explanation. Without this guidance, the process may extend to 7 or 8 minutes, taking precious class time but not providing real additional benefit. With the planning, however, the instructor can make space for an interesting problem, student—student interaction, and feedback from the instructor. Note that there are times that you will want to modify your plan; if students don't understand a key topic, then additional discussion may be helpful.

3. *After class:* Consider providing students an opportunity to solidify and extend understanding.

   The flipped classroom model specifically considers preclass and in-class work, but it doesn't preclude work that students do after class to deepen their understanding. It may be useful to have students continue work that they began in class or to tackle a new but related task (Bruff, 2015). In considering these tasks, however, it's important to consider the #1 pitfall of the flipped classroom.

WHAT ARE PITFALLS TO AVOID?

1. *Too much work.* When instructors report resistance to the flipped classroom, they are often reporting student resistance to a significant increase in the amount of work required for the class. As instructors, we know that increased time on task improves student learning, and the flipped classroom gives us mechanisms to help that happen. It can be wise, however, to be judicious when increasing expectations for student work, both by emphasizing (and demonstrating) the value of additional work to your students and perhaps by taking an iterative approach, such as adding preclass assignments one semester and postclass assignments the next.
2. *Repeating the content from the preclass prep work during an in-class lecture.* Many of us are confident lecturers, and it can be easy to fall back

on this approach when students appear not to have prepared for class or do not engage with the questions or problems we pose. The problem, of course, is that it reinforces behavior we don't want—lack of preparation or participation—and devalues the work of students who did prepare. If students are not engaging before or during class, it may be because of pitfall #3.

3. *Not explaining the rationale to the students.* Students often come to our classes expecting to listen and take notes during a well-organized lecture and to study those notes and the text in the days leading up to an exam. When we violate those expectations, it's important to explain why we are doing so. If we tell our students why we ask them to do preclass preparation and the value that is added from in-class activities, they ascribe more value to the activities as well as greater expertise to the instructor.

4. *Not aligning high-stakes assessments with pre- and in-class work.* Imagine students' dismay if they do preclass work that prepares them to practice applying concepts to case studies in class, followed by additional application questions after class ... and they show up to find an exam that is a series of multiple choice questions demanding memory of discrete facts. That sort of misalignment between instruction and assessment causes a loss of trust that can derail a class. With any instructional method, it's important to align learning goals, activities, and assessments, and that's particularly true for a flipped classroom that demands intensive student engagement.

CONCLUSION

The flipped classroom builds on many effective teaching practices. Importantly, it focuses class time on active learning, often with students working in groups, helping students make connections between new information and existing knowledge to extend their understanding. By giving students a chance to practice key skills with the support of their peers and the instructor, it can also foster a sense of belonging and self-efficacy, two of the key elements that can promote students' motivation and science identity as discussed in Chapter 1. Finally, by explicitly introducing new information before class and then providing opportunities in class to process it, the flipped classroom may also serve as a tool to help students manage the cognitive load of our very complex subjects.

In this section of the book, we are exploring four specific pedagogies that incorporate and build on fundamental teaching and learning approaches. In this chapter, we've considered the flipped classroom; in the next, we turn to theory and practice that can make the most of the educational videos that often power the flipped classroom.

REFERENCES

Abeysekera, L., Dawson, P., 2015. Motivation and cognitive load in the flipped classroom: definition, rationale and a call for research. High. Educ. Res. Dev. 34 (1), 1−14.

Anderson, L.W., Krathwohl, D., 2001. A Taxonomy for Learning, Teaching, and Assessing: A Revision of Bloom's Taxonomy of Educational Objectives. Longman, New York, NY.

Berrett, D., 2012. How 'flipping' the classroom can improve the traditional lecture. The Chronicle of Higher Education, Febraury 19.

Bogost, I., 2013. The condensed classroom. The Atlantic, August 27.

Bransford, J.D., Brown, A.L., Cocking, R.R., 2000. How People Learn: Brain, Mind, Experience, and School. Natl Acad Press, Washington, DC.

Bruff, D., 2015. The Flipped Classroom FAQ. Retrieved from: https://derekbruff.org/?p = 3088 (20.06.18).

Clark, R.C., Nguyen, F., Sweller, J., 2005. Efficiency in Learning: Evidence-Based Guidelines to Manage Cognitive Load. Pfeiffer, San Francisco, CA.

Crouch, C.H., Mazur, E., 2001. Peer instruction: ten years of experience and results. Am. J. Phys. 69, 970−977.

DesLauriers, L., Schelew, E., Wieman, C., 2011. Improved learning in a large-enrollment physics class. Science 332, 862−864.

Eddy, S.L., Hogan, K.A., 2014. Getting under the hood: how and for whom does increasing course structure work? CBE Life Sci. Educ. 13, 453−468.

Fitzpatrick, M., 2012. Classroom lectures go digital. New York Times (Print) June 24.

Gross, D., Pietri Evava, S., Anderson, G., Moyano-Camihort, K., Graham, M.J., 2015. Increased preclass preparation underlies student outcome improvement in the flipped classroom. CBE Life Sci. Educ. 14, 1−8.

Hake, R., 1998. Interactive-engagement versus traditional methods: a six-thousand-student survey of mechanics test data for introductory physics courses. Am. J. Phys. 66, 64−74.

Jensen, J.L., Kummer, T.A., Godoy, P.D., 2015. Improvements from a flipped classroom may simply be the fruits of active learning. CBE Life Sci. Educ. 14, 1−12.

Lage, M.J., Platt, G.J., Treglia, M., 2000. Inverting the classroom: a gateway to creating an inclusive learning environment. J. Econ. Educ. 31, 30−43.

Lockman, K., Haines, S.T., McPherson, M.L., 2017. Improved learning outcomes after flipping a therapeutics module: results of a controlled trial. Acad. Med. Available from: https://doi. org/10.1097/ACM.0000000000001742.

McLean, S., Attardi, S.M., Faden, L., Goldszmidt, M., 2016. Flipped classrooms and student learning: not just surface gains. Adv. Physiol. Educ. 40, 47−55.

Novak, G., Patterson, E.T., Gavrin, A.D., Christian, W., 1999. Just-in-Time Teaching: Blending Active Learning with Web Technology. Prentice Hall, Upper Saddle River, NJ.

Stockwell, B.R., Stockwell, M.S., Cennamo, M., Jiang, E., 2015. Blended learning improves science education. Cell 162, 933−936.

Strayer, J.F., 2012. How learning in an inverted classroom influences cooperation, innovation and task orientation. Learn. Environ. Res. Int. J. 15 (2), 171−193.

Van Vliet, E.A., Winnips, J.C., Brouwer, N., 2015. Flipped-class pedagogy enhances student metacognition and collaborative-learning strategies in higher education but effect does not persist. CBE Life Sci. Educ. 14, 1−10.

Vygotsky, L.S., 1978. Mind in Society. Harvard University Press, Cambridge, MA.

Walvoord, B.E., Anderson, V.J., 1998. Effective Grading: A Tool for Learning and Assessment. Jossey-Bass, San Francisco, CA.

Chapter 10

Using Educational Videos

Video has become an important part of higher education. It is integrated as part of traditional courses, serves as a cornerstone of many blended courses, and is a favored type of resource for students' out-of-class studying. Several meta-analyses have shown that technology can enhance learning (e.g., Schmid et al., 2014), and multiple studies have shown that video, specifically, can be a highly effective educational tool (e.g., Kay, 2012; Allen and Smith, 2012; Lloyd and Robertson, 2012; Rackaway, 2012; Hsin and Cigas, 2013). In order for video to achieve its potential as a productive part of a learning experience, however, it is important for instructors to consider several elements of video design and use. This chapter describes a theoretical framework for considering effective design and use of educational videos as well as specific practices that have been shown to enhance students' learning from video.

WHAT ARE PRINCIPLES FOR EFFECTIVE VIDEO DESIGN AND USE?

Educational videos are usually intended to serve similar functions as a lecture: to explain new content, model ways of thinking, illustrate problem-solving approaches, etc. To consider factors that enhance learning from video, it may therefore be useful to revisit the model of memory formation that we used to consider effective lecture practices. As students watch a video, they encounter an ongoing stream of sensory information through both auditory and visual channels. Students select some of this information to pay attention to in working memory, which has a low capacity and is relatively short-term. If they connect that information to their current understanding of a topic, it promotes encoding of the new information into long-term memory. Thus learning as an interpretive process; new memories form when information is related to what is already known.

When considering educational videos, it can be useful to consider Cognitive Load Theory, which builds on the model of memory formation we have been using throughout this book (Fig. 10.1). Cognitive Load Theory highlights the limited capacity of working memory and emphasizes the role that instructional choices can play in overwhelming working memory and thus limiting learning (Sweller, 1988, 1989, 1994). Alternatively, the theory

Science Teaching Essentials. DOI: https://doi.org/10.1016/B978-0-12-814702-3.00010-X

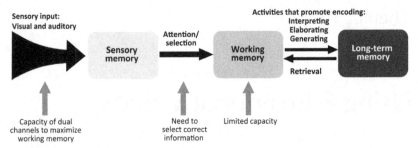

FIGURE 10.1 Cognitive Load Theory and the Cognitive Theory of Multimedia Learning point to elements of memory formation important for learning from video (Sweller, 1988, 1989, 1994; Mayer, 2008). Some sensory input is selected for processing in working memory, which has low capacity and is short term. Connecting information in working memory to existing knowledge leads to encoding in long-term memory. *The model shown is based on the Atkinson–Shiffrin model, supplemented with information from deWinstanley and Bjork* (Atkinson and Shiffrin, 1968; deWinstanley and Bjork, 2002). *Although the model can be useful for making teaching choices, it's important to note that it's highly simplified. In particular, short-term working memory and long-term memory function as a continuum rather than as two distinct entities.*

can point to instructional choices that enhance conditions for understanding and learning.

It may not be obvious why the capacity of working memory, and thus Cognitive Load Theory, is particularly relevant for video. After all, we always encounter an ongoing stream of sensory information, whether we are listening to a lecture, reading a textbook, or doing any other educational activity. With most educational activities, however, there is a high level of human control: the instructor sees confused looks and pauses for clarification, or the reader realizes that she didn't understand a section and goes back to reread. Unless we include tools to encourage this type of human control, however, the sensory-rich nature of video can exceed learners' working memory or can lead them to attend to the wrong elements. Thus, the capacity of working memory may be particularly relevant for this educational tool.

Cognitive Load Theory identifies three components within any learning experience. *Intrinsic cognitive load* is inherent to the subject under study. It is commonly defined as being determined by the degrees of connectivity within the subject. For example, word pairs (e.g., blue = azul) are often given to illustrate a subject with low intrinsic load, whereas grammar is described as a subject with a high intrinsic load due to its many levels of connectivity and conditional relationships. Cognitive Load Theory identifies a second component of a learning experience as the germane load, which is the cognitive activity necessary to reach the desired learning outcome—for example, to make the comparisons, do the analysis, or elucidate the steps necessary to master the lesson. The ultimate goal of these activities is for the learner to incorporate the subject under study into a network of richly connected ideas that allow the learner to use the information. Cognitive Load

Theory identifies a third component of a learning experience as extraneous load, which is cognitive effort that does not help the learner toward the desired learning outcome. It is often characterized as load that arises from a poorly designed lesson (e.g., confusing instructions, extra information).

The Cognitive Theory of Multimedia Learning, articulated by Richard Mayer and colleagues, applies Cognitive Load Theory to multimedia instruction. It emphasizes the role of the dual visual and auditory channels for information acquisition and processing (Mayer and Moreno, 2003) and has been used to investigate design strategies intended to facilitate understanding and integration of new information into existing cognitive structures (Mayer, 2008).

These theories have implications for design and use of educational videos. They suggest that for topics that have high intrinsic load—most topics that students encounter in college science courses—instructors should minimize extraneous load, eliminating unneeded sensory input that doesn't help students reach the learning goal. The theories also suggest that for such complex topics, instructors should design tools to manage intrinsic load. Videos provide rich opportunities to manage intrinsic load: using the auditory and the visual channels to bring complementary information into working memory enhances working memory, and using features that give students control over the progress of the video provides tools for students to manage cognitive load. These theories also suggest that we should optimize germane load, which can be accomplished by highlighting the most important elements for students to select for processing or by including questions and prompts that emphasize what the student should be learning. Thus these theories provide a gateway to identifying practices that can enhance learning from video.

WHAT ARE KEY RECOMMENDATIONS?

There are several practices that have been shown to enhance learning from multimedia instruction, and, in some cases, from video in particular.

1. *Practices that may help manage intrinsic load.* Features that help students manage intrinsic load are powerful means to enhance students' use of and learning from multimedia instruction. These features generally fall into two categories: the use of complementary visual and auditory information to help students understand a complex phenomenon, and the use of features that "chunk" information and give students control.
   - *Matching modality* is the practice of matching the manner in which information is conveyed to the information itself. Every person who has taken a chemistry class knows that it is more effective to have an image of a molecule for reference rather than words alone when discussing oxidation, just as learning about the parts of a flower is much

easier when there is an image to examine. Thus, matching modality is a practice that comes naturally in most science classes. Video provides an opportunity to take this practice a step further, using movement to show relationships or changes over time. Using both the audio/verbal channel and the visual/pictorial channel to convey new information may help students manage intrinsic load by maximizing their working memory. For example, showing an animation of a process on screen while narrating it uses both channels to elucidate the process, thus giving the learner dual and complementary streams of information to highlight features that should be processed in working memory. In contrast, showing the animation while also showing printed text uses only the visual channel and thus overloads this channel and impedes learning (Mayer and Moreno, 2003). In another example, using a "talking head" video to explain a complex process makes productive use only of the verbal channel (because watching the speaker does not convey additional information), whereas a Khan-style tutorial that provides symbolic sketches to illustrate the verbal explanation uses both channels to give complementary information. Using both channels to convey appropriate and complementary information has been shown to increase students' retention and ability to transfer information (Mayer and Moreno, 2003) and to increase student engagement with videos (Thomsen et al., 2014; Guo et al., 2014).

- *Segmenting* is the chunking of information to allow learners to engage with small pieces of new information and to give them control over the flow of new information. As such, it helps students manage intrinsic load, and has been shown to be a powerful way to promote student learning from video (Ibrahim et al., 2012). Segmenting may also increase germane load by emphasizing the structure of the information.

Segmenting can be accomplished in several ways. One of the easiest, and perhaps most productive, is to keep videos short and focused. Guo and colleagues examined the length of time students watched streaming videos within four edX MOOCs, analyzing results from 6.9 million video watching sessions (Guo et al., 2014). They observed that the median engagement time for videos less than six minutes long was close to 100%—that is, students tended to watch the whole video (although they did observe significant outliers; see the paper for more complete information). As videos lengthened, however, student engagement dropped off, such that the median engagement time with 9−12 minutes videos was ∼50% and the median engagement time with 12−40 minutes videos was ∼20%. In fact, the maximum median engagement time for a video of any length was six minutes. Making videos longer than 6−9 minutes is therefore likely to be wasted effort. While this study examined whether students watched videos rather than whether they learned from them, watching is a prerequisite for learning—and thus segmenting your lesson into short videos may help your students engage and learn.

Another mechanism to segment your videos and help students manage intrinsic cognitive load is to use interactive features that give students control. These can be "chapters" that are essentially labeled segments within a video, or can be click-forward pauses, both features that are readily introduced with most video-editing software, such as Camtasia, Screen-Cast-o-Matic, or H5P. Zhang and colleagues compared the impact of interactive and noninteractive video on students' learning in a computer science course (Zhang et al., 2006). Students who were able to control movement through the video, selecting important sections to review and moving backwards when desired, demonstrated better achievement of learning outcomes and greater satisfaction. Alternatively, or in addition, it can be helpful to integrate questions into the video. Tools like H5P can allow instructors to incorporate questions directly into video and to give feedback based on student response.

2. *Practices that may help optimize germane load.* Watching a video can be a passive process. We often allow video to flow over us while we are actually thinking about something entirely different. Even when we are actively engaged, we may not correctly identify the key elements in a new-to-us complex process. Features that help students identify the information to which they should be paying attention and the cognitive activity in which they should be engaged—and perhaps more importantly, give them some structure and some reason to engage in that cognitive activity—can enhance the germane load of the learning experience and promote student learning.

- *Signaling*, also known as cueing (deKoning et al., 2009), is the use of on-screen text or symbols to highlight important information. For example, signaling may be provided by the appearance of two or three key words (e.g., Mayer and Johnson, 2008; Ibrahim et al., 2012), a change in color or contrast (e.g., deKoning et al., 2009), or a symbol that draws attention to a region of a screen (e.g., an arrow; deKoning et al., 2009). By highlighting the key information, it helps direct learner attention, thus targeting particular elements of the video for processing in the working memory. This practice can increase germane load by emphasizing to what students should pay attention as well as the organization of and connections within the information. Mayer and Moreno (2003) and deKoning et al. (2009) have shown that this approach improves students' ability to retain and transfer new knowledge from animations, and Ibrahim et al. (2012) have shown that these effects extend to video.

- *Using guiding questions* helps students understand to what they should be paying attention and the kinds of thinking they should do while watching a video. Lawson and colleagues examined the impact of guiding questions on students' learning from a video about social psychology in an introductory psychology class (Lawson et al., 2006). They had students in some sections of the course watch the video with no special instructions, while students in other sections of the course were provided with eight guiding questions to consider while watching. The students who answered the guiding questions while watching the video scored significantly higher on a later test.

- *Integrating questions into the video* can also help students understand what they should be thinking about when watching a video, but these questions also provide an opportunity for immediate feedback that can help students assess their understanding. Tools like H5P can allow instructors to incorporate questions directly into video and to give feedback based on student response, providing formative feedback that can be useful for both the student and the instructor. Vural (2013) compared the effect of video with embedded questions to interactive video without embedded questions in preservice teachers, finding that the embedded questions improved the students' performance on subsequent quizzes.

- *Making video part of a larger homework assignment* provides students with direct and concrete information about what they should be learning from the video, and it may also increase student motivation to use the video to enhance their understanding. Faizan Zubair developed videos that were embedded in a larger homework assignment in Paul Laibinis' Chemical Engineering class at Vanderbilt University. He found that students valued the videos and that they improved students' understanding of difficult concepts when compared to a semester when the videos were not used in conjunction with the homework (https://vanderbilt.edu//bold/docs/chbe-162-chemical-engineering-thermodynamics/).

3. *Practices that reduce extraneous load.* When we teach, we have endless opportunities to add information that is peripheral to what students are actually learning, such as stories, interesting applications, or cool demonstrations. Video provides additional opportunities to add features like music and entertaining backgrounds. These additions can be useful, serving as hooks that promote engagement and enhance the value of the lesson for students. When students are learning about something complex and challenging, however, this type of extraneous information should be minimized to avoid overwhelming students' working memory.
   - *Weeding* is the elimination of interesting but extraneous information from the video, that is, information that does not contribute to the learning goal. For example, music, complex backgrounds, or extra

features within an animation require students to judge whether to pay attention to them, which increases extraneous load and can reduce learning. Importantly, information that increases extraneous load changes as the learner moves from novice toward expert status. That is, information that may be extraneous for a novice learner may actually be helpful for a more expert-like learner, while information that is essential for a novice may serve as an already-known distraction for an expert. Thus, it's important that the instructor consider her students when weeding educational videos, including information that is necessary for their processing but eliminating information that they don't need to reach the learning goal and that may overload their working memory. Ibrahim has shown that this treatment can improve retention and transfer of new information from video (Ibrahim et al., 2012). Importantly, this does not mean that instructors should eliminate those interesting "hooks" to help promote student engagement with and value for the subject. Instead, those stories, applications, or demonstrations should be segmented so that students can engage with them before (and maybe also after) grappling with the hard work of understanding the complex subject.

- *Eliminating redundancy* can help students learn (Sweller, 1988). When students are presented with the same information in two different formats, and either can be understood without the other, it increases the cognitive load of the lesson without providing any benefit. Instructors may include this sort of redundancy in an effort to match their students' preferred learning style, but multiple studies have shown that it is better to identify the mode most appropriate for the information and that redundant information reduces student learning (Sweller, 1988; Mayer, 2008).

  What does this mean in practice for educational videos? It means that providing printed text to accompany a diagram or animation, particularly when accompanied by the type of narration usually provided in a video, increases cognitive load and tends to reduce student learning (Mayer, 2008). Video, which relies heavily on complementary auditory and visual information, can readily avoid the redundancy effect.

4. *Practices that help students engage with video.* Most of the practices that have been shown to impact students' learning from video and other multimedia lessons are related to managing the cognitive load of the lesson. There are a few practices that appear to have more to do with students' perception of narration style.

- *Using a conversational style* for narration enhances student engagement with video. Called the personalization principle by Richard Mayer, the use of conversational rather than formal language during multimedia instruction has been shown to have a large effect on students' learning, perhaps because a conversational style encourages students to develop sense of social partnership with the narrator that leads to greater engagement and effort (Mayer, 2008).

- *Speaking quickly and with enthusiasm* also enhances student engagement with video. In their study examining student engagement with MOOC videos, Guo and colleagues observed that student engagement was dependent on the narrator's speaking rate, with student engagement increasing as speaking rate increased (Guo et al., 2014). It can be tempting for video narrators to speak slowly to help ensure that students grasp important ideas, but including in-video questions, "chapters," and speed control can give students control over this feature—and increasing narrator speed appears to promote student interest.

CONCLUSION

Cognitive Load Theory and the Cognitive Theory of Multimedia Learning point us to several good practices that enhance learning from video. The overarching theme is that educational video requires particular attention to cognitive load. By using the auditory and visual channels in complementary ways and by segmenting lessons into digestible chunks, instructors can create videos that help students learn content that is inherently complex. Eliminating extraneous information enhances this effect, helping ensure that students can focus on the key features of the lesson. In addition, providing tools that help students identify important elements and do the cognitive processing that will help them learn further enhances learning. Finally, using a conversational and enthusiastic style when creating video narration can help students stay engaged. Videos created or used with attention to these elements can prove to be an engaging tool to add to your teaching.

In this section of the book, we are exploring four specific pedagogies that incorporate and build on fundamental teaching and learning approaches. In the final chapter for this section, we turn to a pedagogy that can meld science instructors' identities as teachers and researchers: incorporating authentic research experiences into credit-bearing courses.

REFERENCES

Allen, W.A., Smith, A.R., 2012. Effects of video podcasting on psychomotor and cognitive performance, attitudes and study behavior of student physical therapists. Innov. Educ. Teach. Int. 49, 401–414.

Atkinson, R.C., Shiffrin, R.M., 1968. Human memory: a proposed system and its control processes. In: Spence, K.W., Spence, J.T. (Eds.), The Psychology of Learning and Motivation, vol. 2. Academic Press, New York, NY, pp. 89–195.

deKoning, B., Tabbers, H., Rikers, R., Paas, F., 2009. Towards a framework for attention cueing in instructional animations: guidelines for research and design. Educ. Psychol. Rev. 21, 113–140.

deWinstanley, P.A., Bjork, R.A., 2002. Successful lecturing: presenting information in ways that engage effective processing. New Dir. Teach. Learn. 89, 19−31.

Guo, P.J., Kim, J., Robin, R., 2014. How video production affects student engagement: an empirical study of MOOC videos. ACM Conference on Learning at Scale (L@S 2014). http:// groups.csail.mit.edu/uid/other-pubs/las2014-pguo-engagement.pdf.

Hsin, W.J., Cigas, J., 2013. Short videos improve student learning in online education. J. Comput. Sci. Coll. 28, 253−259.

Ibrahim, M., Antonenko, P.D., Greenwood, C.M., Wheeler, D., 2012. Effects of segmenting, signaling, and weeding on learning from educational video. Learn. Media Technol. 37, 220−235.

Kay, R.H., 2012. Exploring the use of video podcasts in education: a comprehensive review of the literature. Comput. Hum. Behav. 28, 820−831.

Lawson, T.J., Bodle, J.H., Houlette, M.A., Haubner, R.R., 2006. Guiding questions enhance student learning from educational videos. Teach. Psychol. 33, 31−33.

Lloyd, S.A., Robertson, C.L., 2012. Screencast tutorials enhance student learning of statistics. Teach. Psychol. 39, 67−71.

Mayer, R.E., 2008. Applying the science of learning: evidence-based principles for the design of multimedia instruction. Cogn. Instr. 19, 177−213.

Mayer, R.E., Johnson, C.I., 2008. Revising the redundancy principle in multimedia learning. J. Educ. Psychol. 100, 380−386.

Mayer, R.E., Moreno, R., 2003. Nine ways to reduce cognitive load in multimedia learning. Educ. Psychol. 38, 43−52.

Rackaway, C., 2012. Video killed the textbook star? Use of multimedia supplements to enhance student learning. J. Polit. Sci. Educ. 8, 189−200.

Schmid, R.F., Bernard, R.M., Borokhovski, E., Tamim, R.M., Abrami, P.C., Surkes, M.A., et al., 2014. The effects of technology use in postsecondary education: a meta-analysis of classroom applications. Comput. Educ. 72, 271−291.

Sweller, J., 1988. Cognitive load during problem solving: effects on learning. Cogn. Sci. 12, 257−285.

Sweller, J., 1989. Cognitive technology: some procedures for facilitating learning and problem-solving in mathematics and science. J. Educ. Psychol. 81, 457−466.

Sweller, J., 1994. Cognitive load theory, learning difficulty, and instructional design. Learn. Instr. 4, 295−312.

Thomsen, A., Bridgstock, R., Willems, C., 2014. 'Teachers flipping out' beyond the online lecture: maximising the educational potential of video. J. Learn. Des. 7, 67−78.

Vural, O.F., 2013. The impact of a question-embedded video-based learning tool on e-learning. Educ. Sci.: Theory Pract. 13, 1315−1323.

Zhang, D., Zhou, L., Briggs, R.O., Nunamaker Jr., J.F., 2006. Instructional video in e-learning: assessing the impact of interactive video on learning effectiveness. Inform. Manag. 43, 15−27.

Chapter 11

Incorporating Research Into Courses

with Faith Rovenolt

There is long-standing and widespread support for educational initiatives that introduce undergraduates to scientific research (e.g., Karukstis and Elgren, 2007; Healey 2013; Alberts 2009). Research experiences have been shown to lead to student-reported gains in general skills (e.g., oral, visual, and written communication) as well as more specific research-associated skills (e.g., research design, hypothesis formation, data analysis) (Seymour et al., 2004; Lopatto, 2006; Laursen et al., 2010). Traditionally, the apprentice model has been the primary mechanism for introducing students to research, with undergraduates joining a research group and receiving one-on-one guidance from a faculty mentor or another group member (Van Dyke et al., 2017). This model can be highly effective, producing not only student-reported gains but also student persistence in STEM (Eagan et al., 2013; Schultz et al, 2011), and it is deeply embedded in the culture of science. This model also, however, limits the number of students who can engage in research experiences and tends to be more accessible to students who are familiar with the research enterprise and who are further along in their undergraduate career, which can disproportionally affect minority or underrepresented students (Bangera and Brownell, 2014).

There is therefore an increasing emphasis on opportunities for students to engage in research in credit-bearing courses, extending the benefits of research experiences to a larger and more diverse group (PCAST, 2012; National Academies of Sciences, Engineering, and Medicine, 2015, 2017). These course-based research experiences, variously called course-based undergraduate research experiences (CUREs), course-based research experiences (CREs), discovery-based research courses, and authentic laboratory undergraduate research experiences (ALUREs), can provide many of the same benefits students derive from traditional undergraduate research experiences, such as increased content knowledge and analytical skills (Jordan

Science Teaching Essentials. DOI: https://doi.org/10.1016/B978-0-12-814702-3.00011-1

143

et al., 2014; Corwin et al., 2015) and persistence in science (Rodenbusch et al., 2016; Shaffer et al., 2014).

This chapter describes important characteristics of course-based research experiences and provides examples that range from national programs to single courses developed by individual instructors. We end the chapter by offering practical suggestions to help instructors design an effective course-based research experience.

WHAT ARE IMPORTANT CHARACTERISTICS OF A COURSE-BASED RESEARCH EXPERIENCE?

A 2017 report from the National Academies of Sciences, Engineering, and Medicine identifies three types of goals for students participating in undergraduate research experiences (National Academies of Sciences, Engineering, and Medicine, 2017). These goals are to:

- increase participation and retention of STEM students;
- integrate students into STEM culture;
- promote the development of STEM disciplinary knowledge and the use of STEM research practices (Box 11.1).

Each of these goals is multifaceted and complex, and collectively they represent much of what we hope our students gain throughout our curriculum. In many ways, however, research experiences are the most authentic representation of what it means to be a scientist, and thus community goals for these experiences are high.

Of course, all lab experiences—traditional, inquiry-based, or research-based—are intended to help students move toward at least one of these goals. Further, analyzing scientific literature can, on its own, provide benefits (Box 11.2). When thinking about whether and how to incorporate research into your own course, it may be helpful to compare CUREs with traditional course labs, inquiry-based labs, and the apprentice-model research experience (URE) (see Fig. 11.1). While students in all of these lab experiences

---

BOX 11.1 Disciplinary Research Practices

- defining questions and problems
- developing and using models
- planning and interpreting data
- using mathematics and computational thinking
- constructing explanations and arguing from evidence
- communicating questions, methods, results, and conclusions

BOX 11.2 Analyzing Science Communication as a Gateway to Understanding Research

One of the major benefits of introducing students to scientific research is to develop students' understanding of the nature and development of scientific knowledge (Lopatto, 2003; National Research Council, 2003). Several groups have demonstrated that careful analysis of common forms of science communication can provide this benefit without having physically to engage in a lab, thereby providing a gateway for students into research.

- *Using the research talk to help students understand the research process and how knowledge is constructed* (Clark et al., 2009). Banerjee and colleagues sought a low resource mechanism to introduce first- and second-year undergraduates to the process of scientific discovery. The solution they developed:

Introduction	An invited speaker gives an hour-long, full-scale research seminar
Weeks 1–5	The course instructor leads students through the seminar in 5–10 chunks, helping students identify hypotheses, explore the experimental approaches, and analyze the data presented.
	Students complete problems sets outside of class that ask them to formulate hypotheses, propose experiments, and suggest future directions for research.
Question and answer wrap-up	The invited speaker returns for an hour-long question-and-answer session.

Each student completes two modules during the course. Banerjee and colleagues used the Classroom Undergraduate Research survey to determine whether this approach achieved the goals of undergraduate research (Lopatto, 2007). They found that students completing this course exhibited self-reported learning greater than students completing summer research experiences in several categories, notably understanding the research process, understanding supporting evidence, and understanding how new knowledge is constructed. Thus the model Banerjee and colleagues propose may provide a readily adaptable tool for helping students gain an understanding of the process of research.

- *Using the CREATE model to analyze scientific literature and understand the process of discovery* (Gottesman and Hoskins, 2013; Hoskins et al., 2007). Hoskins and colleagues developed the CREATE model of analyzing the scientific literature to give undergraduates an understanding of how scientific knowledge is generated and how research projects progress over time. CREATE focuses on a sequence of articles that reports a single line of research from one laboratory. Students receive each article in sections, and are asked to work through the data as if they had generated it themselves. The steps they follow:

Consider	Students read the introduction and construct a concept map to highlight the essential concepts and questions in the article.

(Continued)

BOX 11.2 (Continued)

Read	Students read the Methods and Results section and then "work backward" from the results to the experiments, creating a visual representation (a sketch or diagram) of each experiment, labeling figures, and writing descriptive titles for the diagrams and figures.
Elucidate the hypothesis	Students identify the specific hypothesis tested in each experiment.
Analyze and interpret the data	Students analyze the data, comparing control and experimental panels, relating the results to the hypothesis, evaluating the data, and noting questions for the authors.
Think of the next Experiment	Each student diagrams two experiments that they think should be done next. These are reviewed by the class in a process that mimics the peer-review process.

Students analyze the papers in the series sequentially, discussing each paper before moving on to the next, related paper. This allows the students to see if the experiments they proposed are those selected by the authors, both helping them understand the role of creativity in scientific progress and creating a "lab meeting" atmosphere in the class.

Hoskins and colleagues have implemented this approach in both upper-level and introductory classes. Its use in an upper-level course led to gains in students' content integration, critical thinking ability, and self-assessed learning gains. It was adapted for use in a course for first-year students, who showed a significant increase in critical thinking ability, experimental design ability, and self-rated abilities such as decoding literature, thinking like a scientist, and understanding research in context. Thus, the CREATE approach to analyzing primary literature may also provide a readily adaptable tool for helping students gain an understanding of the process of research.

Traditional "Cookbook" Verification Labs	Inquiry Labs	CUREs (Course-based undergraduate research experience)	UREs (Apprentice-model undergraduate research experience)
• Students may use mathematics and computational modeling • Students may construct explanations and argue from evidence • Students do not design questions or experimental approach	• Students engage in a range of research practices depending on learning goals, including experimental design, data collection and analysis, and constructing explanations and arguments • Students may also identify a research question and engage in other research practices, such as reporting to colleagues in the class	• Students engage in a range of practices as appropriate to the research question • Students understand that they are generating results to contribute to the understanding of the scientific field about a larger question • Students design some element of the project, collect and analyze data, construct explanations, and communicate results	• Students engage in a range of practices as appropriate to the research question • Students understand that they are generating results to contribute to the understanding of the scientific field about a larger question • Students collect and analyze data, construct explanations, and communicate results; they may design some element of the project
• Questions and answers are known in advance	• The answers are unknown to the students but may be known or of limited interest to the scientific community • The emphasis is on promoting student interest and the student learning process	• Collaboration with other students is emphasized • Student design may be more limited than in inquiry labs • The importance of data quality is greater than in inquiry labs	• Student design may be more limited than in inquiry labs • The importance of data quality is very high

FIGURE 11.1   Common features of undergraduate lab experiences.

engage in multiple scientific practices, inquiry labs, and CUREs tend to ask students to engage in a greater range of scientific practices than traditional cookbook labs (Weaver et al., 2008). Inquiry labs and CUREs tend to emphasize different practices, however, reflecting some different underlying goals and assumptions. Inquiry labs are focused on the student learning process and seek to engage students' interest by giving them control over multiple elements of the project, ranging from experimental design to identification of the research question. An underlying assumption is that student autonomy promotes deep engagement with, and persistent interest in, science. An underlying assumption of CUREs, on the other hand, is that becoming a part of a larger scientific community and contributing to a research question of interest to that community promotes student interest and persistence. For these reasons, students may have less freedom to design questions and experiments in CUREs than in inquiry labs, but there is greater emphasis on contributing to a growing body of knowledge. While students should have the opportunity to design some element of the project, there are often constraints associated with addressing an "authentic" research question: the larger question, as well as the tools and methods used to produce meaningful data, are typically defined by the instructor. In these ways, course-based research experiences may have more in common with research apprenticeships.

The CURE description in Fig. 11.1 represents an emerging consensus around elements that allow CUREs to fit community definitions of a research experience (Dolan, 2016; Auchincloss et al., 2014; Heemstra et al., 2017). These key elements can be summarized as

- *Discovery:* Students have the opportunity to make discoveries by collecting and analyzing novel data and producing results that are new to them and to the scientific community.
- *Relevance:* Students pursue research questions that build on existing knowledge and that are of interest to a larger scientific community.
- *Scientific practice:* Students engage in multiple scientific practices, such as reading literature, designing some aspect of the project, analyzing data, making interpretations, communicating results, and framing work in the larger body of knowledge.
- *Collaboration:* Students work with colleagues and instructors, often to tackle a shared problem. The forms the collaboration can take varies among disciplines and projects but provides students an opportunity to understand the benefits of many people working on a complex problem.
- *Communication:* Students communicate their results publicly to an audience with a vested interest in the work. The opportunity for students to share their results outside the classroom and contribute to a growing body of knowledge is an important factor for research experiences.

As you are designing your course, it is worth considering how these elements connect to the larger goals for research experiences. In other words, instructors may find it helpful to consider which mechanisms may link these elements of discovery, relevance, scientific practice, collaboration, and communication to increased participation and retention of STEM students, integration of students into STEM culture, and promotion of STEM disciplinary knowledge, and practice. Corwin and colleagues have proposed a model that can help instructors and researchers examine these relationships (Corwin et al., 2015). Using situated learning theory as a guide, the model relates common student activities in CUREs to outcomes that have been observed in at least two student populations in at least two studies. For example, the model suggests both complementary and converging pathways by which collecting novel data (discovery) may enhance science identity and persistence in science (Fig. 11.2), building on short-term outcomes of increased technical

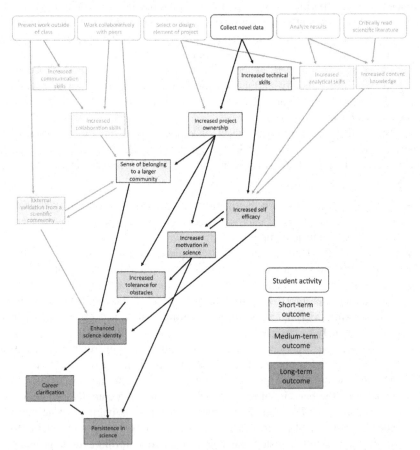

FIGURE 11.2 Model relating common student activities in CUREs to outcomes observed in at least two populations in two studies. Source: *Modified from Corwin, L.A., Graham, M.J., Dolan, E.L., 2015. Modeling course-based undergraduate research experiences: an agenda for future research and evaluation. CBE Life Sci. Educ. 14(1), es1.*

skills, project ownership, and sense of belonging to achieve medium-term outcomes of increased motivation and selfefficacy, ultimately contributing to the desired long-term outcomes. Analogous pathways can be drawn for the other key features of CUREs. Thus, this model may help instructors make reasoned choices about the types of activities in which students will engage—from reading the literature to designing methods—to achieve their learning goals. It may also serve as a guide as instructors consider the outcomes they wish to measure to determine the impact of the experience.

Of course, many aspects of students' activities will be shaped by their particular research, and some of the most important questions about designing course-based research experiences have to do with choosing an appropriate and sustainable project for students to pursue. Instructors have used a variety of ways to set up these experiences, from joining a national network to designing their own. The next section presents four models for course-based research experiences, with examples from each.

WHAT FORMS CAN CURES TAKE?

There are four types of CUREs that have been reported in the literature: national programs with a common research goal, national programs that provide access to technology or instrumentation, programs that support varied course-based research experiences at a particular institution, and specific, instructor-generated courses (modified from Dolan, 2016). The examples described below are intended to serve as links, templates, or inspiration for instructors choosing ways to incorporate research experiences into their courses, depending on the instructor's goals and constraints.

National programs with a common research goal focus on large questions that benefit from many contributors. They provide an opportunity for students and faculty to contribute to understanding a larger problem, and offer students a great example of the collaborative nature of modern science. They also provide significant support for instructors, including training, centralized resources, and a community of like-minded teacher-scholars.

- *The Small World Initiative*, developed at Yale in 2012, provides opportunities for students to identify novel antibiotic leads by isolating local soil bacteria, testing those bacteria against clinically-relevant microorganisms, and characterizing bacteria that exhibit antibiotic activity (www.smallworldinitiative.org). The program provides instructional materials, instructor training, access to online tools, and ongoing support and has been implemented at more than 270 institutions in 14 countries, ranging from research-intensive universities to community colleges. Small World Initiative projects have been implemented in microbiology, biology for majors and nonmajors, and cell and molecular biology courses, and provide opportunities to expand projects to fit the goals of particular groups of students and instructors (Davis et al., 2017). Student grades and pre- and postcourse measurements revealed that students participating in the Small World Initiative at Florida Atlantic University demonstrated higher grades and greater increases in critical thinking skills than students in a control, cookbook style lab (Caruso et al., 2016).

- *The Genomics Education Partnership (GEP)*, spearheaded by Sarah Elgin, is a collaboration between multiple colleges and universities and the Department of Biology and Genome Center of Washington University (gep.wustl.edu; Shaffer et al., 2010). Undergraduate researchers improve the sequence of and annotate various *Drosophila* genomes to allow analysis of the differences between heterochromatin and euchromatin. Single students or small groups of students can work on sequence improvement, designing primers to target areas of interest, specifying the type of sequencing to be done, and analyzing the sequence data. Alternatively, they can work on an annotation project, identifying putative start and stop sites and intron and exon boundaries to produce a gene model. Projects have been adapted for stand-alone courses as well as modules within existing lab courses. The program provides support for faculty through online resources and a week-long workshop that introduces them to the project. Pre- and postcourse quizzes have demonstrated an increase in content knowledge, while a postcourse survey has indicated professional and learning gains similar to students in summer research programs (Lopatto et al., 2008; Shaffer et al., 2010).
- *The Science Education Alliance—Phage Hunters Advancing Genomics and Evolutionary Science (SEA—PHAGES)* project, led by Graham Hatfull, centers around a two-semester course in which students isolate and characterize bacteriophage (that is, viruses that infect bacteria) from local soils, extract and purify DNA for sequencing, and annotate the genome (Hatfull et al., 2006). Participation in the program produced gains in content knowledge, math skills, and experimental design and data interpretation skills at Gonzaga University (Staub et al., 2016). As of 2014, more than 4800 students at 73 institutions had contributed to the project (Jordan et al., 2014).

National programs that provide access to technology or instrumentation also support instructors in engaging their students in research experiences. In addition to providing access to technology or instrumentation, they may also provide training or research modules that instructors can adopt.

- *The Genome Consortium for Active Teaching using Next-Generation Sequencing (GCAT-SEEK)* provides faculty and students with resources to use massively parallel sequencing as a tool for promoting undergraduate research experiences (Buonaccorsi et al., 2014). Members of the consortium share expertise both on the technical elements of next-generation sequencing and on pedagogical and assessment choices relevant to incorporating research into a credit-bearing lab, providing faculty development workshops, access to computational resources, and opportunities for shared data, discounts, and sequencing efforts. The program adopts the model used by the Genome Consortium for Active Teaching developed by Malcolm Campbell at Davidson College, which has been active in supporting course-based research experiences since 1999.

- *The Center for Authentic Science Practice in Education (CASPiE)*, a collaboration among chemistry departments at Purdue University and several other universities and colleges, (Weaver et al., 2006) combined elements of programs with common research goals and programs that provide access to instrumentation. Specifically, scientists design modules based on their own research that can be adopted by first- and second-year chemistry lab courses. Using peer-led team learning approaches, students in the courses design experiments and generate results that can be used by the module author in his or her research program. The module author can have varied levels of involvement with the courses that have adopted the module, from minimal to significant. Course instructors may develop their own modules and share them for use at other institutions as well. There is a central support structure, with CASPiE faculty and staff providing guidance on module development, helping with module testing, and providing a remote instrument resource (www.purdue.edu/discovery-park/caspie). Research modules appropriate for General Chemistry and Organic Chemistry courses have been developed and range from questions about the effects of food processing on phytochemical antioxidants, to design of a potential antiviral drug candidate, to characterization of semiconducting films' use for solar energy conversion. Russell and Weaver demonstrated that students participating in research as part of this program showed greater understanding of essential elements of scientific research concepts (e.g., controls, repeatability, and the importance of unanticipated results) than did students in inquiry-based or traditional labs (Russell and Weaver, 2011).

Programs that support varied course-based research experiences at a particular institution. A few institutions have developed programs intended to support course-based research across disciplines at a particular institution. These programs provide administrative support for developing, implementing, and assessing CUREs related to research interests of individual faculty members.

- *The Freshman Research Initiative* at the University of Texas at Austin (https://cns.utexas.edu/fri) provides first-year students in the College of Natural Sciences the opportunity to engage in course-based research from their first semester. Students take a sequence of three courses. In the first course, they learn to read scientific literature and design and execute at least one scientific investigation. In addition, they are matched to a research "stream" in one of more than 25 areas within a range of disciplines, including astronomy, biology, biochemistry, bioinformatics, chemistry, computer science, and physics, each incorporating research projects developed by faculty members. In the second course, students learn about the overall research goals for their stream as well as essential concepts and skills related to the research. They also begin to contribute to the research efforts of the group. Finally, in the third course, students focus on contributing to the research efforts of the stream, often proposing and completing a research subproject (Rodenbusch et al., 2016). More than 900 students participate in the FRI each year, and completion of all three courses in the program has been demonstrated to increase the likelihood of completing a STEM degree and graduating within six years (Rodenbusch et al., 2016). Further, Ghanem and colleagues describe five FRI chemistry projects that accomplished American Chemical Society accreditation objectives and resulted in publishable results (Ghanem et al., 2017).

- *Vertically Integrated Projects* at Georgia Tech (www.vip.gatech.edu) provide opportunities for students to engage in course-based research projects based on faculty research interests for up to three years. The projects and the teams that pursue them are multidisciplinary; project descriptions explicitly draw attention to the ways that different disciplinary perspectives can contribute to the research. Each team has students from sophomores to PhD students as well as a faculty lead. The projects are varied, ranging from themes such as Academic Resilience to Bio-Inspired Network Dynamics and Geomechanics to Lightning from the Edge of Space, but the courses that students take have common elements, such as research notebooks, peer evaluations, and student-designed rubrics for self-evaluation. The program originated in engineering (Coyle et al., 2006) and has been adapted for use in other disciplines and at multiple institutions (e.g., Coyle et al., 2015; Choi and Kim, 2017; Sonnenberg-Klein et al., 2017).

Specific courses. Most CUREs are developed by individual faculty, based on their research interests, the educational needs of their students, and the resources available to them. Science education has a rich history of innovative instructors seeking to engage their students in research, both with and without publishing these efforts. The following examples are chosen to highlight published examples of course-based research experiences in different science disciplines, but discipline-specific education journals can provide many other wonderful examples to serve as inspiration or templates.

- *Biology.* The greatest number of published course-based research experiences are in the biological sciences, and an investigation of the CUREnet database provides multiple examples in biology subfields from bioinformatics to ecology. Flaherty and colleagues describe a course-based research experience for wildlife biology students at the University of Wyoming (Flaherty et al., 2017). Junior and senior undergraduate students investigate effects of climate change on chipmunk populations, designing their own specific questions and research methods during each semester of the course. For example, students have investigated the effects of shrub encroachment and tree mortality on chipmunk population dynamics, leading to reports for the Wyoming Game and Fish Department as well as presentation in the departmental seminar series. Participation produced benefits similar to those associated with apprentice-style research experiences, improving students understanding of disciplinary concepts, the scientific process, and confidence in their science knowledge. It also clarified career choice for many students.

- *Chemistry.* Many courses in chemistry incorporate opportunities for students to do research (see e.g., Heemstra et al., 2017). Williams and Reddish describe a course-based research experience for an upper-level physical chemistry course at Emory University (Williams and Reddish, 2018). Students used molecular dynamics simulations, isothermal titration calorimetry, and stopped-flow kinetics to study ligand binding to human serum albumin, an experience that increased students' confidence in their ability to carry out a research project and sense of project ownership. May and colleagues also observed an increase in students' confidence in their research skills and attitudes toward chemistry in a lab course in which students investigated pH and ion content of Arctic snow in the context of an environmental chemistry question (May et al., 2018).

- *Geoscience.* Course-based research is a fairly standard feature in geoscience, particularly in capstone courses, but it is becoming more integrated throughout the curriculum (Mogk and Goodwin, 2012; Ryan, 2014). In the Council of Undergraduate Research's *Developing and sustaining a research-supportive curriculum: A compendium of successful practices,* Karukstis and Elgren (2007) describe their systematic efforts to incorporate controlled research experiences into the Geology curriculum to complement their capstone research projects. One of the major curricular adjustments they made to reach this goal was the design of a Research Methods course with a shifting focus. Each year, the course is designed around a central theme or location, which allows students to develop individual projects but also promotes group collaboration. For example, in 2002, the central theme was field-based research at San Onofre State Beach, CA, a theme that led to one subgroup of students collaborating to collect data relevant to rock deformation—but then individually collecting additional data to complete individual projects. The course includes written and oral presentations, with an iteratively revised journal-style manuscript. Pre- and postcourse surveys indicated that the course provided benefits similar to that students experienced in summer research internships. In addition, informal assessment indicated that the faculty perceived an increase in the quality of senior theses. Finally, they observed an increase in the number of students who were active in research, both in and out of class, noting an increase from 17% to 30% of students presenting a conference abstract and an increase from 8% to 28% participating in research prior to senior year.

- *Physics.* There are fewer published reports of labs that instructors characterize as course-based research experiences in physics, although programs like the Freshman Research Initiative and Vertically Integrated Projects have multiple examples of physics-based research projects. Several reports do address the transformation of lab courses to incorporate more authentic scientific practices. For example, Lewandowski and Finkelstein describe the redesign of a junior level electronics course required for physics majors at University of Colorado, Boulder (Lewandowski and Finkelstein, 2015). The course was redesigned to align student activities with current practice in electronics research, to incorporate more design and application activities, and to increase students' project ownership. The redesign incorporated activities to enhance student preparation for labs, to emphasize use of class time for analyzing as well as collecting data, and to increase accountability and engagement in student-designed projects. Preliminary results indicated that students in the transformed course were more likely to identify a discrepancy in a model they had developed and to refine that model, a key practice in physics research.

PRACTICAL SUGGESTIONS

The examples of course-based research experiences described here and elsewhere (Dolan, 2016; Van Dyke et al., 2017) can serve as guides in developing or adopting a research experience for your course. Collectively, they offer guidelines for technical choices and pedagogical considerations related to CUREs.

Technical Considerations

- It's important to choose an overarching project that can be addressed with reliable methods that undergraduates can easily learn and implement. What those methods are will vary among disciplines and projects, but finicky protocols are likely to lead to disappointing results for projects that involve multiple students who are learning a new technique.
- CUREs typically have to fit within an undergraduate schedule, where students are in the lab for 3–4 hours once or twice a week. Projects that have multiple discrete steps and that can be paused and restarted as needed are a good fit for most undergraduate calendars.
- CUREs that allow students to design and "own" small and complementary pieces of larger projects can be highly satisfying to students,

allowing them to develop a sense of ownership while also emphasizing the value of multiple hands and multiple minds collaborating to answer a larger question.

- It can be helpful to encourage students to build redundancy into the project design—perhaps with two groups completing the same experiment—to emphasize the importance of replication and to serve as a tool for quality control.

Pedagogical Considerations

- As with any learning experience, it's critical to begin your CURE design by identifying your learning goals for your students. By focusing on those learning goals, you can choose student activities that not only allow students to take the steps needed to move their research forward but also to reflect on the reasons for and benefits of these steps.
- Students can find collaboration challenging. It's a critical element of modern research, so taking time to talk about the roles it can play—from collaborative project design, to bringing complementary expertise, to peer review—is worthwhile. It can also be helpful to provide students an opportunity to talk about goals and norms for their collaboration. Specific ideas are provided in Chapter 5.
- Students can also find the frequent failure and ambiguity of research to be challenging. Addressing "failure" both at the beginning and as it arises can help them place it in the context of how the field builds knowledge.
- Sharing results with the broader community provides students with a sense of belonging to a larger scientific community and can help build their identity as scientists. Identifying a mechanism for doing this that is satisfying and meaningful to your students is key. For example, some students may value less formal but more immediate avenues for sharing results, such as contributing to a database or an online library, more than later authorship on a research article (Wiley and Stover, 2014).

CONCLUSION

Course-based research experiences have the potential to expand the benefits of research to a larger and more diverse set of students, increasing students' experience with and understanding of the ways science builds knowledge and their interest in pursuing a science degree. These experiences also have particular value for instructors, providing an opportunity for faculty to integrate the teaching and research aspects of their professional identities. Faculty who have taught courses that incorporate research report that teaching them is "rewarding," "fulfilling," and "intellectually stimulating," while also offering opportunities to collect data for research programs, to pilot research projects, and to obtain grant funding (Shortlidge et al., 2017). In addition, CUREs can provide valuable teaching opportunities for graduate students and postdoctoral scientists (Cascella and Jez, 2018). While challenging to implement, course-based research experiences provide a wealth of benefits for students and faculty that make them worth considering for your teaching.

In this section of the book, we have explored four specific pedagogies that incorporate and build on fundamental teaching and learning approaches. In the final section of the book, we turn to practices that help ensure that our assessments are a fair and reliable measure of the skills and knowledge we want our students to gain, first considering good practices in test writing and concluding with rubric development and use.

REFERENCES

Alberts, B., 2009. Redefining science education. Science 323, 437.

Auchincloss, L.C., Laursen, S.L., Branchaw, J.L., Eagan, K., Graham, M., Hanauer, D.I., et al., 2014. Assessment of course-based undergraduate research experiences: a meeting report. CBE Life Sci. Educ. 13, 29–40.

Bangera, G., Brownell, S.E., 2014. Course-based undergraduate research experiences can make scientific research more inclusive. CBE Life Sci. Educ. 13 (4), 602–606.

Buonaccorsi, V., Peterson, M., Lamendella, G., Newman, J., Trun, N., Tobin, T., et al., 2014. Vision and change through the genome consortium for active teaching using next-generation sequencing (GCAT–SEEK). CBE Life Sci. Educ. 13, 1–2.

Caruso, J.P., Israel, N., Rowland, K., Lovelace, M.J., Saunders, M.J., 2016. Citizen science: the small world initiative improved lecture grades and California critical thinking skills test scores of nonscience major students at Florida Atlantic University. J. Microbiol. Biol. Educ. 17 (1), 156.

Cascella, B., Jez, J.M., 2018. Beyond the teaching assistantship: CURE leadership as a training platform for future faculty. J. Chem. Educ. 95 (1), 3–6.

Choi, J.E., Kim, H., 2017. Vertically integrated projects (VIP) at Inha University: the effect of convergence project education on learning satisfaction. Teaching, Assessment, and Learning for Engineering (TALE), 2017 IEEE 6th International Conference on. IEEE, pp. 436–443.

Clark, I.E., Romero-Calderon, R., Olson, J.M., Jaworkski, L., Lopatto, D., Banerjee, U., 2009. "Deconstructing" scientific research: a practical and scalable tool to provide evidence-based science instruction. PLOS Biol. 7, e1000264.

Corwin, L.A., Graham, M.J., Dolan, E.L., 2015. Modeling course-based undergraduate research experiences: an agenda for future research and evaluation. CBE Life Sci. Educ. 14 (1), es1.

Coyle, E.J., Allebach, J.P., and Garton Krueger, J. (June 2006). The vertically integrated projects (VIP) program in ECE at Purdue: fully integrating undergraduate education and graduate research. Paper Presented at the Proceedings of the 2006 ASEE Annual Conference and Exposition, Chicago, IL. pp. 11.1336.1–11.1336.16. Available at: https://peer.asee.org/1421.

Coyle, E.J., Krogmeier, J.V., Abler, R.T., Johnson, A., Marshall, S., Gilchrist, B.E., 2015. The vertically integrated projects (VIP) program: leveraging faculty research interests to transform undergraduate STEM education. TransformingInstitutions: Undergraduate STEM Education for the 21st Century. Purdue University Press, pp. 223–234.

Davis, E., Sloan, T., Aurelius, K., Barbour, A., Bodey, E., Clark, B., et al., 2017. Antibiotic discovery throughout the small world initiative: a molecular strategy to identify biosynthetic gene clusters involved in antagonistic activity. Microbiol. Open 6 (3).

Dolan, E.L., 2016. Course-based undergraduate research experiences: current knowledge and future directions. National Research Council Commissioned Paper, Washington, DC.

Eagan, M.K., Hurtado, S., Chang, M.J., Garcia, G.A., Herrera, F.A., Garibay, J.C., 2013. Making a difference in science education the impact of undergraduate research programs. Am. Educ. Res. J. 50, 683–713.

Ghanem, E., Long, S.R., Rodenbusch, S.E., Shear, R.I., Beckham, J.T., Procko, K., et al., 2017. Teaching through research: alignment of core chemistry competencies and skills within a multidisciplinary research framework. J. Chem. Educ. 95 (2), 248–258.

Flaherty, E.A., Walker, S.M., Forrester, J.H., Ben-David, M., 2017. Effects of course-based undergraduate research experiences (CURE) on wildlife students. Wildl. Soc. Bull. 41, 701–711.

Gottesman, A.J., Hoskins, S.G., 2013. CREATE cornerstone: introduction to scientific thinking, a new course for STEM-interested freshmen, demystifies scientific thinking through analysis of scientific literature. CBE Life Sci. Educ. 12, 59–72.

Hatfull, G.F., Pedulla, M.L., Jacobs-Sera, D., Cichon, P.M., Foley, A., Ford, M.E., et al., 2006. Exploring the mycobacteriophage metaproteome: phage genomics as an educational platform. PLoS Genet. 2 (6), e92.

Healey, M., 2013. Linking research and teaching: a selected bibliography. Available at: www.mickhealey.co.uk/resources.

Heemstra, J.M., Waterman, R., Antos, J.M., Beuning, P.J., Bur, S.K., Columbus, L., et al., 2017. Throwing away the cookbook: implementing course-based undergraduate research experiences (CUREs) in chemistry. Educational and Outreach Projects from the Cottrell Scholars Collaborative Undergraduate and Graduate Education, vol. 1. American Chemical Society, pp. 33–63.

Hoskins, S.G., Stevens, L.M., Nehm, R.H., 2007. Selective use of the primary literature transforms the classroom into a virtual laboratory. Genetics 176 (3), 1381–1389.

Jordan, T.C., Burnett, S.H., Carson, S., Caruso, S.M., Clase, K., DeJong, R.J., et al., 2014. A broadly implementable research course in phage discovery and genomics for first-year undergraduate students. MBio 5 (1), e01051-13.

Karukstis, K.K., Elgren, T.E., 2007. Developing and Sustaining a Research-Supportive Curriculum: A Compendium of Successful Practices. Council on Undergraduate Research.

Laursen, S., Hunter, A., Seymour, E., Thiry, H., Melton, G., 2010. Undergraduate Research in the Sciences: Engaging Students in Real Science. Jossey-Bass, San Francisco, CA.

Lewandowski, H.J., & Finkelstein, N. (2015). Redesigning a junior-level electronics course to support engagement in scientific practices. In: Churukian, J., Ding, L. (Eds.), 2015 Physics Education Research Conference Proceedings.

Lopatto, D., 2003. The essential features of undergraduate research. Counc. Undergrad. Res. Quart. 23, 139–142.

Lopatto, D., 2006. Undergraduate research as a catalyst for liberal learning. Peer Rev. 8 (1), 22–25.

Lopatto, D., 2007. Undergraduate research experiences support science career decisions and active learning. CBE Life Sci. Educ. 6, 297–306.

Lopatto, D., Alvarez, C., Barnard, D., Chandrasekaran, C., Chung, H.-M., Du, C., et al., 2008. Undergraduate research. Science 322, 684–685.

May, N.W., McNamara, S.M., Wang, S., Kolesar, K.R., Vernon, J., Wolfe, J.P., et al., 2018. Polar plunge: semester-long snow chemistry research in the general chemistry laboratory. J. Chem. Educ. 95 (4), 543–552.

Mogk, D.W., Goodwin, C., 2012. Learning in the field: synthesis of research on thinking and learning in the geosciences. Geol. Soc. Am. Spec. Pap. 486, 131–163.

National Academies of Sciences, Engineering, and Medicine, 2015. Integrating Discovery-Based Research Into the Undergraduate Curriculum: Report of a Convocation. National Academies Press.

National Academies of Sciences, Engineering, and Medicine, 2017. Undergraduate Research Experiences for STEM Students: Successes, Challenges, and Opportunities. The National Academies Press, Washington, DC. Available from: https://doi.org/10.17226/24622.

National Research Council, 2003. BIO 2010: Transforming Undergraduate Science Education for Future Research Biologists. National Academies Press, Washington, DC.

President's Council of Advisors on Science and Technology (PCAST) (2012). Engage to excel: producing one million additional college graduates with degrees in science, technology, engineering, and mathematics. Retrieved from: www.whitehouse.gov/sites/default/files/microsites/ostp/pcast-engage-to-excel-final_2-25-12.pdf.

Rodenbusch, S.E., Hernandez, P.R., Simmons, S.L., Dolan, E.L., 2016. Early engagement in course-based research increases graduation rates and completion of science, engineering, and mathematics degrees. CBE Life Sci. Educ. 15 (2), ar20.

Russell, C.B., Weaver, G.C., 2011. A comparative study of traditional, inquiry-based, and research-based laboratory curricula: impacts on understanding of the nature of science. Chem. Educ. Res. Pract 12, 57–67.

Ryan, J.G., 2014. Supporting the transition from geoscience student to researcher through classroom investigations using remotely operable analytical instruments. Geoscience Research and Education. Springer, Dordrecht, pp. 149–162.

Schultz, P.W., Hernandez, P.R., Woodcock, A., Estrada, M., Chance, R.C., Aguilar, M., et al., 2011. Patching the pipeline reducing educational disparities in the sciences through minority training programs. Educ. Eval. Pol. Anal. 33, 95–114.

Seymour, E., Hunter, A.B., Laursen, S.L., DeAntoni, T., 2004. Establishing the benefits of undergraduate research for undergraduates in the sciences: first findings from a three-year study. Sci. Educ. 88, 493–534.

Shaffer, C.D., Alvarez, C., Bailey, C., Barnard, D., Bhalla, S., Chandrasekaran, C., et al., 2010. The genomics education partnership: successful integration of research into laboratory classes at a diverse group of undergraduate institutions. CBE Life Sci. Educ. 9, 55–69.

Shaffer, C.D., Alvarez, C.J., Bednarski, A.E., Dunbar, D., Goodman, A.L., Reinke, C., et al., 2014. A course-based research experience: how benefits change with increased investment in instructional time. CBE Life Sci. Educ. 13 (1), 111–130.

Shortlidge, E.E., Bangera, G., Brownell, S.E., 2017. Each to their own CURE: faculty who teach course-based undergraduate research experiences report why you too should teach a CURE. J. Microbiol. Biol. Educ. 18 (2).

Sonnenberg-Klein, J., Abler, R.T., Coyle, E.J., 2017. Multidisciplinary vertically integrated teams: social network analysis of peer evaluations for vertically integrated projects (VIP) program teams.

Staub, N.L., Poxleitner, M., Braley, A., Smith-Flores, H., Pribbenow, C.M., Jaworski, L., et al., 2016. Scaling up: adaptation of a phage-hunting course to increase participation of first-year students in research. CBE Life Sci. Educ. 15 (2), ar13.

Van Dyke, A.R., Gatazka, D.H., Hanania, M.M., 2017. Innovations in undergraduate chemical biology education. ACS Chem. Biol. 13 (1), 26–35.

Weaver, G., Wink, D., Varma-Nelson, P., Lytle, F., Morris, R., Fornes, W., et al., 2006. Developing a new model to provide first and second-year undergraduates with chemistry research experience: early findings of the Center for Authentic Science Practice in Education (CASPIE). Chem. Educator 11, 125–129.

Weaver, G.C., Russell, C.B., Wink, D.J., 2008. Inquiry-based and research-based laboratory pedagogies in undergraduate science. Nat. Chem. Biol. 4 (10), 577.

Wiley, E.A., Stover, N.A., 2014. Immediate dissemination of student discoveries to a model organism database enhances classroom-based research experiences. CBE Life Sci. Educ. 13, 131–138.

Williams, L.D., Reddish, M.J., 2018. Integrating primary research into the teaching lab: benefits and impacts of a one-semester CURE for physical chemistry. J. Chem. Educ. Available from: https://doi.org/10.1021/acs.jchemed.7b00855Article ASAP.

Section IV

Fair and Transparent Grading Practices

Chapter 12

Writing Exams: Good Practice for Writing Multiple Choice and Constructed Response Test Questions

Giving exams can be one of the most stressful parts of teaching. The challenge begins with the writing: how do you ensure that your exam fairly covers course content? What question constructions help uncover student knowledge? The challenge continues during grading: how do you maximize the probability that exams are graded equitably across the class? And the exams you write also have larger implications: what role do the assessments in your course, including the types of exams students take, play in determining how students study and how they learn? This chapter addresses these questions, first examining important principles to consider when writing exams and then turning to specific recommendations for exam planning and question writing. We end the chapter considering test expectancy, or the role that students' expectations about an exam plays in determining what and how they learn.

WHAT ARE IMPORTANT PRINCIPLES IN EXAM CONSTRUCTION?

There are a few key principles to consider when thinking about exams, notably validity and reliability. These concepts are similar to the ideas of accuracy and precision. Validity is the extent to which an exam, or an exam question, measures what we want it to measure, and reliability is the extent to which an exam or an exam question can be expected to gives the same result when used repeatedly. Here, we examine the ideas of validity and reliability and testing elements that can undermine them.

Validity

Validity is the extent to which an exam (or an exam question) measures the knowledge and skills it purports to measure (Cheser Jacobs and Chase,

Science Teaching Essentials. DOI: https://doi.org/10.1016/B978-0-12-814702-3.00012-3

1992). Does the question focus on whether students can demonstrate appropriate understanding of the course content, or does it primarily measure some other skill or knowledge set?

On first consideration, it seems obvious that exams would target course content in an appropriate way. The instructor is the exam-writer, so he or she knows what should be on the exam! In practice, however, it can be easy for exam questions primarily to measure skills or even knowledge that are not the main focus of the course. This can happen in several ways.

- Multiple choice questions that have a lot of extraneous information or that ask students to distinguish among A&D, AB&D but not C, etc., require a robust working memory to hold all the relevant information in mind simultaneously. The question does not really distinguish among students who know the course content and those who don't, but instead who know the course content and has test-taking skills and working memory that allow them to parse the questions.
- Alternatively, multiple choice questions can give clues to the right answer, helping out students who know how to read those clues but not really measuring understanding of course content. For example, test-wise students know that answers including "always" and "never" are unlikely correct responses, and that instructors will often write the longest option for the correct answer as they try to ensure that it is exactly right.
- Both constructed-response and multiple-choice questions can target trivial knowledge that is not at the heart of the course and isn't really what the instructor cares about—these questions are often easy to write, and they are able to distinguish among students in the class, but they are not valid measures of students' appropriate understanding of the course content.
- Questions using words that are not part of the course and that may be unfamiliar to nonnative speakers or first-generation college students are unfair measures for the course. For example, the seemingly simple words *net* and *gross* are familiar to most native English speakers and can be used to ask students to take into account both investment and output for a given process, but they're unfamiliar to many nonnative speakers and thus produce an unfair question.
- Constructed response questions suffer from some of their own threats to validity. Specifically, writing skills may boost the grade students get on a constructed response item, making a mediocre answer appear stronger than it is.

Thus, there are many ways to unintentionally construct an exam or an exam item that is not a valid measure for your course, so it's worthwhile to be mindful of concept throughout exam construction and grading.

Reliability

A second concept that can be important to exams is reliability. In essence, reliability is the consistency or repeatability of your exam for measuring students' knowledge. That is, if you could give two students with equivalent knowledge and skills the same exam in the same way, reliability is a measure of the probability that they would receive the same score.

For instructor-generated classroom exams, there are two major factors to consider with regard to reliability: what impact does guessing have on students' scores, and what impact does grading variability have? Guessing is primarily a threat for multiple choice and other kinds of selected response items, and it's exaggerated for some types of selected response questions (notably, true/false). Conversely, variation in grading is primarily a threat for constructed-response questions, where placement in a stack of papers can impact the scores students receive. For example, a mediocre constructed response that is read immediately after a really poor one is likely to receive a higher score than *the same mediocre response* that is read immediately after an excellent example.

Because exams are measuring student choices that represent knowledge and skills, validity and reliability are related. For example, an exam question that is very wordy and causes a student to trip up in different ways is not reliable, and is also not a valid measure of the students' understanding of the question content. Happily, there are relatively simple, concrete suggestions that can be used to improve both validity and reliability.

WHAT ARE GENERAL RECOMMENDATIONS FOR GETTING STARTED?

The first and perhaps most important recommendation for constructing exams is to consider what should be on it before you start writing. What content should the exam cover? What skills should students demonstrate? What fraction of your test should require higher order thinking, and what fraction should measure basic knowledge?

To clarify the answers to these questions, it is good practice to construct a table of specifications (aka, a test blueprint) to guide your test writing. A table of specifications characterizes the distribution of points across content areas and types of thinking you want the test to target. It can be easiest to start with a distribution of points across content areas as demonstrated in Table 12.1.

TABLE 12.1 Example Content Distribution for Test

Topic	Percentage of Test Points (%)
Protein structure	10
Enzyme catalytic mechanisms	30
Enzyme kinetics	30
Enzyme regulation	30

TABLE 12.2 Example Content Distribution and Desired Ways of Demonstrating Knowledge for Test

	Remember (%)	Explain/ Describe (%)	Apply/ Calculate (%)	Analyze/ Evaluate (%)
Protein structure	–	5	–	5
Enzyme catalytic mechanisms	5	10	–	10
Enzyme kinetics	–	10	20	5
Enzyme regulation	5	5	10	10
Total percentages	10	30	30	30

The table of specifications can then be expanded to include the types of thinking the instructor wants the students to do within each content area (Table 12.2).

Ways of demonstrating knowledge will vary by discipline and instructor, and higher-level skills may encompass lower-level skills. There is therefore no set recommendation for the types of thinking the table should include; it's simply a tool to help the test-maker be sure to target appropriate skills. You may find it most appropriate to have two levels of questions, those requiring higher-level thinking and those primarily requiring memory, or you may want to include a finer-grained distribution as illustrated in Table 12.2. In either case, it can be helpful to consider Bloom's taxonomy and associated verbs that characterize different levels of thinking, a few of which are shown here.

- *Remember:* define, list, repeat, state.
- *Understand:* classify, describe, discuss, explain, identify, select.
- *Apply:* execute, solve, demonstrate, interpret, sketch.
- *Analyze:* organize, compare, contrast, examine, relate, predict.
- *Evaluate:* argue, defend, support, critique, appraise.
- *Create:* design, construct, develop, formulate.

WHAT ARE RECOMMENDATIONS FOR MULTIPLE CHOICE AND OTHER SELECTED RESPONSE QUESTIONS?

After constructing a table of specifications, we turn to writing the questions themselves. For many of us, the questions include multiple choice and other types of selected response items. Instructors often bemoan the need to include multiple choice items, but when aligned with course goals, they can be valuable measures of student understanding. They are not prone to subjective grading and thus tend to be more objective and often more reliable measures of student learning than other types of questions. They tend to take less time to answer than constructed response questions and thus allow a greater representation of course material, with the possibility of examining different levels of student understanding for a given topic. They are, of course, challenging to write, and some constructions can advantage the "test-wise" while others target unintended skills (e.g., rapid reading comprehension or working memory). The recommendations below, drawn largely from a comprehensive review by Haladyna et al. (2002) and the National Board of Medical Examiners' guidelines (2016), can help instructors avoid problematic constructions and thus focus their items on targeted knowledge and skills.

Item Format

Selected response items can be one-best-answer (i.e., traditional multiple choice), true-false, multiple true-false (MTF; aka, choose all correct), or matching (Fig. 12.1). Haladyna and colleagues' concluded that all of these item formats can be valid components of classroom exams, although the National Board of Medical Examiners (NBME) recommends one-best-answer items as

Multiple Choice Item Type	Item Example	Recommendation
One best answer	Which subatomic particle is found *outside* of the nucleus of an atom? ⌐ Stem A. Electron B. Neutron ⊢ Alternatives C. Proton	✓✓
Complex multiple choice	What conditions increase overall activity of the citric acid cycle? I. high concentration of NAD^+ II. high concentration of Ca^{2+} III. high concentration of AMP IV. high concentration of citrate A. I only B. I, III C. I, III, IV D. I, IV E. I, II, III, IV	Not recommended
Multiple true–false	What conditions increase overall activity of the citric acid cycle? Next to each option write *True* or *False*. I. high concentration of NAD^+ II. high concentration of Ca^{2+} III. high concentration of AMP IV. high concentration of citrate	✓

FIGURE 12.1   Types of selected response items.

the least ambiguous and most trustworthy type of items. The NBME notes that because students choose the *best* answer to the item, distractors can be partially true, requiring students to exercise judgement about the most correct answer. The NBME recommends avoiding MTF items because there can be ambiguity about the degree to which a given statement is "true" or "false," and, as a consequence, test-writers may tend to focus these items on lower-order thinking skills (i.e., straight recall). When constructed to have unambiguously true or false alternatives and aligned with the goals of the class, however, these items can be useful exam components.

Both Haladyna and colleagues and the NBME support using item sets, or multiple questions associated with a single scenario or dataset, and note that this type of construction can be efficient for the test-writer and can be effective at testing higher-order thinking skills. When using item sets, it's important to write questions in the set to be independent, avoiding constructions where one answer depends on a previous answer.

There is one type of selected response item that instructors should absolutely avoid: the complex multiple choice item. These items require students to choose one best answer from choices such as A; B; C; A and B only; B and C only; A and C only; A, B and C. Because these items have flaws that reduce both reliability and validity, they are not useful exam components.

Guidelines for Writing the Stem

There are several recommendations for writing item stems.

- The stem should contain the central idea of the item, and should be should be clear and meaningful by itself. The NBME describes this as "front-loading" items and recommends checking to see if it's possible to "cover the options" and still answer the question.
- The stem should not contain extraneous information, or "window dressing." Extraneous information can make an item more a test of reading speed and comprehension rather than the targeted knowledge and can reduce both reliability and validity (Haladyna and Downing, 1989).
- The stem can be a question or an incomplete statement.
- It's best to avoid negative stem constructions, such as "which molecule is not ...?" or "all species except ..." If a stem is negatively stated, it's important to highlight the negative word in some way (such as with capitalization or bolding).

Guidelines for Writing Alternatives

There are also several recommendations for writing multiple-choice alternatives (Fig. 12.2).

Characteristics	Example
Convergence	Insulin leads to the dephosphorylation and activation of FBPase-2. Which of the following would be a direct effect of this activation? A. **Activation** of fatty acid synthesis. B. **Activation** of glycogen synthesis. C. **Activation** of glycolysis. D. **Activation** of gluconeogenesis. E. Inhibition of gluconeogenesis. Alternatives converge on "activation," suggesting it is part of the correct answer.
Paired answers	Insulin leads to the dephosphorylation and activation of FBPase-2. Which of the following would be a direct effect of this activation? A. Activation of fatty acid synthesis. B. Activation of glycogen synthesis. C. Activation of glycolysis. D. **Activation of gluconeogenesis.** E. **Inhibition of gluconeogenesis.** Paired choices suggest one is correct.
Grammatical cues	In mitochondria that have been treated with the ATP synthase blocker oligomycin, O_2 consumption can be restored by administration of _____. A. Ammonium. B. Greater activity within the rest of the cell. C. Paranitrophenol. Grammar is inconsistent with stem, making this alternative implausible.
Answers are different lengths	When performing a Grignard reaction, why is it important to ensure there is *no* water present? A. Grignard reagents irreversibly react with water, causing reaction yield to be lower. B. Magnesium complexes with water. C. Water may react with the alkyl group. One answer is longer, a cue that this answer is correct.
Question wording cues right answer	The process of an atom gaining or losing an electron to form an **ion** is referred to as _____. A. Ionization. B. Oxidation. C. Reduction. D. Vaporization. The word ion in the stem cues students to the correct alternative.

FIGURE 12.2 Cues that help the test-wise: constructions that point to the correct answer.

- All alternatives should be plausible. Implausible options are quickly eliminated and serve no purpose. Haladyna and colleagues recommend writing as many plausible alternatives as possible, but note that three alternatives are sufficient in most instances (Haladyna et al., 2002). Common student errors serve as the best source for incorrect alternatives.
- Alternatives should be stated in clear and concise language. Wordy alternatives test reading skill more than the targeted learning outcomes and reduce the validity of the item.

- Alternatives should have parallel construction and similar length. When one alternative is longer than the others, it tips off test-wise students, who recognize that the instructor may be trying to make that alternative completely correct.
- Alternatives should be free from logical clues about the correct choice. Many of our students are experienced test-takers, and they can be adept at mining the information in alternatives to narrow down their options and choose a correct answer without full knowledge. There are several types of logical cues that help these test-wise students described here and illustrated in Fig. 12.2.
  - The use of absolutes, such as "always" and "never," within alternatives is a strong hint to test-wise students that these are incorrect answers. Our students know that we are careful about accuracy, and there are few situations in which absolutes provide the best answers.
  - Paired alternatives can be logical cues that one of those choices is correct. When using paired alternatives, it may be best to use two or more pairs to avoid unintentional hints to students.
  - Convergence within alternatives can serve as a cue to the test-wise. If a key word is present in several alternatives, it suggests that the correct answer should include this word.
- It's best to avoid "all of the above" and "none of the above" as alternatives. Both of these options allow students with partial knowledge to judge the best answer, reducing the validity and reliability of the item (Haladyna and Downing, 1989).
- Alternatives should be presented in a logical order, such as alphabetical or numerical, to alleviate unintentional biases toward a particular position.

WHAT ARE RECOMMENDATIONS FOR CONSTRUCTED RESPONSE QUESTIONS?

Constructed response items can also be highly effective test components. They can allow students to articulate their reasoning and to show greater depth of knowledge and organization of knowledge than selected response items allow. They are also effective at revealing incomplete student understanding, although they are less effective at revealing misconceptions than selected response items (Hubbard et al., 2017). Hogan and Murphy provide a few recommendations to maximize the effectiveness of constructed-response items (Hogan and Murphy, 2007).

- Be transparent about the point values of the item or the time students should allot to answering the question. This can be written on the test next to the item.
- Define the question or task clearly, using verbs that ensure students know what they're supposed to do. For example, the verbs *compare, sketch,*

and *predict* point to clear actions that students should take in constructing their answer.

- Have a colleague or teaching assistant review the constructed response item.
- Be aware that providing options (e.g., answer 2 of the following 3 questions) reduces the reliability of the tests for comparability purposes.
- Develop a rubric or other scoring strategy, such as a checklist, *when you write the item.* This helps ensure that the item is testing the skills and knowledge that you intend, and it helps ensure consistent grading, which is one of greatest challenges for constructed response items. See the examples in Tables 12.3 and 12.4.
- Consider using more constructed-response questions with less time allocated for each rather than using fewer constructed response questions with more time allocated for each. This can allow greater coverage of course material and thus increase the validity of the exam.

Hogan and Murphy (2007) also provide recommendations for grading constructed-response items to increase validity and reliability.

TABLE 12.3 Example Checklist for Scoring Constructed Response Item

Present ? (Check)	Criteria Student Discusses...
	Observations they made during the investigation
	Data they collected
	The approach they used for collecting data
	Future directions and changes

TABLE 12.4 Example Rubric for Scoring a Constructed Response Item

Criteria	Exceeds Expectations (3 Points)	Meets Expectations (2 Points)
Claim	Clear statement of claim	Claim mentioned but not clearly described
Evidence	Clear statement of evidence to support claim	Evidence mentioned but either not clearly linked to claim or superficial in nature
Justification	Clear description of concepts that explain how evidence supports their claim	Relevant concepts mentioned but not clearly or accurately related to evidence or claim

- Score constructed response items without knowing the identities of the examinees. Some instructors have their students write their names on the back of the test to accomplish this.
- If there are multiple constructed-response items, score one item at a time for each examinee before moving on to the next item. This ensures more consistent grading across items.
- Shuffle tests between questions to ensure maximum reliability and objectivity in scoring. This reduces "neighbor" effects, helping ensure that no one exam is graded immediately after an excellent or a very poor example on repeated questions.
- If multiple people are grading the same question, perform a standardization to ensure that all people are grading in a uniform manner.

HOW CAN YOU BUILD ON STUDENTS' TEST EXPECTANCY?

When planning the role of exams in your course, it may be useful to consider an idea called *test expectancy*. Many sources suggest that the types of assessments in our courses shape students' approaches to learning as well as what they learn, including what they learn to do, how deeply they learn it, and how persistent the learning is. This idea is not new; seminal work from the 1970s suggested that summative assessment within a course strongly influences students' learning (Becker et al., 1968; Snyder, 1971; Joughin, 2010). Recent work has fleshed out this general observation in college science classes, asking particularly how format and level of exams influences students study approach and knowledge. Specifically, Jensen and colleagues varied the types of quizzes and unit tests students took in two sections of a nonmajors general biology course taught in the same way by the same instructor, giving one section high-level, primarily MC tests that required application, evaluation, and analysis, and giving the other section memory-oriented tests (Jensen et al., 2014). Students in both sections took the same final exam, which consisted of half high-level and half low-level MC questions. Students who had taken high-level quizzes and tests throughout the semester performed better on both types of questions on the final exam. Thus, this study suggested that expecting high-level questions on an exam improved students' memory and their ability to perform the high-level thinking.

Importantly, the cues instructors give students about studying appear to matter as well. Thiede and colleagues found that telling students to expect a comprehension test and to focus on self-explanation during study improved students' understanding and performance in an undergraduate psychology course (Wiley et al., 2016). Further, Stanger-Hall (2012) found that students who expected an exam composed of both MC and constructed response items exhibited more cognitively active study behaviors as well as better performance on the final exam, even though ~30% of the MC questions on unit tests required higher-level thinking. She postulated that students in the introductory class expected MC questions to be low level (although this was not consistently true), and that they therefore used less-engaged study strategies when expecting this type of test. Thus, consistently using in-class, quiz, and test questions that require higher level thinking, and emphasizing the importance of self-explanation and reasoning, may improve your students learning at all levels. Further, writing tests with both selected-response and constructed-response questions may also influence their study habits and performance.

CONCLUSION

In brief, there are several fairly straightforward practices we can adopt to help improve our exam validity and reliability. We can start by planning the percentage of the exam to address particular types of thinking and content. We can construct multiple choice items to avoid unnecessary complexity as well as unintentional cues to the right answer. We can develop tools to help grading our constructed response items be more consistent and fair. Finally, it can be helpful to use students' test expectancy to prompt the kinds of learning that you hope to see in your class.

To help us further consider practices that help ensure that our assessments are a fair and reliable measure of the skills and knowledge we want our students to gain, we turn to rubric development and use.

REFERENCES

Becker, H.S., Geer, B., Hughes, E.C., 1968. Making the Grade: The Academic Side of College Life. Transaction, New Brunswick, NJ (Republished in 1995).

Cheser Jacobs, L., Chase, C.I., 1992. Developing and Using Tests Effectively: A Guide for Faculty. Jossey-Bass Publishers, San Francisco.

Haladyna, T.M., Downing, S.M., 1989. Validity of a taxonomy of multiple-choice item-writing rules. Appl. Meas. Educ. 2 (1), 51−78.

Haladyna, T.M., Downing, S.M., Rodriguez, M.C., 2002. A review of multiple choice item-writing guidelines for classroom assessment. Appl. Meas. Educ. 15 (3), 309−334.

Hogan, T.P., Murphy, G., 2007. Recommendations for preparing and scoring constructed-response items: What the experts say. Appl. Meas. Educ. 20 (4), 427−441.

Hubbard, J.K., Potts, M.A., Couch, B.A., 2017. How question types reveal student thinking: an experimental comparison of multiple-true-false and free-response formats. CBE Life Sci. 16 (2), 1−13.

Jensen, J.L., McDaniel, M.A., Woodard, S.M., Kummer, T.A., 2014. Teaching to the test...or testing to teach: exams requiring higher order thinking skills encourage greater conceptual understanding. Educ. Psychol. Rev. 26, 307−329.

Joughin, G., 2010. The hidden curriculum revisited: a critical review of research into the influence of summative assessment on learning. Assess. Eval. High. Educ. 35, 335−345.

National Board of Medical Examiners, 2016. In: Paniagua, M.A., Swygert, K.A. (Eds.), Constructing Written Test Questions for the Basic and Clinical Sciences. National Board of Medical Examiners.

Snyder, B.R., 1971. The Hidden Curriculum. Knopf, New York, NY.

Stanger-Hall, K., 2012. Multiple-choice exams: an obstacle for higher-level thinking in introductory science classes. CBE Life Sci. Educ. 11, 294–306.

Wiley, J., Griffin, T.D., Jaeger, A.J., Jarosz, A.F., Cushen, P.J., Thiede, K.W., 2016. Improving metacomprehension accuracy in an undergraduate course context. J. Exp. Psychol.: Appl. 22 (4), 393–405. Available from: https://doi.org/10.1037/xap0000096.

Chapter 13

Rubrics: Tools to Make Grading More Fair and Efficient

Science faculty almost universally strive to be objective, fair, and consistent. It's an essential element of our research, and it informs how we approach our classes. We also want our students to have opportunities to do open-ended work that lets them practice and demonstrate more scientific skills and knowledge than we can get at with an exam. Some tension can arise from these two goals: how do we help ensure that we are objective, fair, and consistent when we are evaluating work for which there is not a single correct answer? Rubrics provide one tool that can help us align these goals, giving us a means to make our grading more consistent and transparent. They also have the potential to improve students' learning experience and to make our grading more efficient.

The key to realizing these benefits is to identify the kinds of rubrics that work for your learning goals, your students, and your assessment style. This chapter briefly explores what rubrics are and the benefits they can offer before providing practical recommendations for how to construct and use rubrics that are a good fit for your needs.

WHAT ARE THEY?

Rubrics describe the criteria that will be used to judge a student's work as well as some detail about performance levels. Within that general definition, there are many variations.

Holistic rubrics provide descriptions of work that correspond to different performance levels without delineating multiple criteria. For example, the holistic rubric shown in Table 13.1 describes excellent, acceptable, and unacceptable contributions to a course discussion board. All the characteristics of excellent work are clustered together in a single description, with no attempt to break them out into different components. Each performance level can correspond to a particular grade.

Analytic rubrics delineate the criteria by which a students' work will be judged. They also provide some description of different performance levels for each of these criteria, although the detail of those descriptions can vary.

Science Teaching Essentials. DOI: https://doi.org/10.1016/B978-0-12-814702-3.00013-5

TABLE 13.1 Example Holistic Rubric

Excellent	Discussion post responds to the prompt or to another student's post, offers information from class or from the text as well as information from other sources, and provides a question that can extend discussion.
Acceptable	Discussion post responds to the prompt or to another student's post and offers information from class or from the text.
Unacceptable	Discussion post does not respond to the prompt or to another student's post or fails to offer information from class or from the text.

TABLE 13.2 Example Analytic Rubric

	Excellent	Acceptable	Unacceptable
Relevance	Discussion post responds to the prompt or to another student's post.	Discussion post responds to the prompt or to another student's post.	Discussion post off topic; does not respond to the prompt nor to another student's post.
Use of evidence	Post offers information from class or from the text. Post offers information from other scientific sources.	Post offers information from class or from the text.	Post provides only student's thoughts, does not provide evidence from scientific sources.
Engages others	Post provides a question that can extend discussion.	Post includes a question, but the question has simple yes/no answer and so cannot extend discussion.	Post does not seek to extend discussion by offering a question.

For example, Table 13.2 provides an analytic rubric for the same discussion board assignment viewed in Table 13.1. Because analytic rubrics break out criteria, they tend to provide more detail about what produces different levels of performance within each category.

Single-point rubrics are a variation on the analytic rubric. Like analytic rubrics, they delineate the criteria by which students work will be judged. Rather than describing different performance levels, however, they describe acceptable work and provide space for the instructor to provide feedback on

TABLE 13.3 Example Single-Point Rubric

Areas for Improvement	Criteria and Description of Acceptable Work	Areas Exceeding Standards
	Relevance: discussion post responds to the prompt or to another student's post.	
	Use of evidence: post offers information from class or from the text.	
	Engages others: post includes a question, but the question has simple yes/no answer and so cannot extend discussion.	

ways the work exceeds or falls short of that level. Table 13.3 provides an example of a single-point rubric for the same discussion post assignment viewed in Tables 13.1 and 13.2.

Each of these types of rubrics can be valuable, and each has its limitations. The holistic rubric allows for the most rapid grading and can be particularly valuable for low-stakes, repeated assignments. The descriptions of different performance levels provide targets that students can consider in completing the assignment. Holistic rubrics don't, however, provide detailed feedback. For higher stakes assignments, it may be useful to provide the type of more detailed feedback that an analytic rubric or a single-point rubric facilitates. Analytic rubrics have the advantage of describing different performance levels for each criterion, giving instructors the opportunity to provide specific feedback on different elements of students' work with a few swipes of a pen (or computer mouse). They are, however, challenging to write, and they can feel overwhelming to both instructors and students who are reading them. Single-point rubrics require less cognitive load to interpret and provide an opportunity for both positive and corrective feedback on specific elements of the students' work. Because they do not describe exemplary work, however, students may be less likely to include the elements that would move their work from acceptable to excellent. Thus there is no single best way to write or to use a rubric. The key is to consider how to choose a rubric that suits your style, your students, and your assignment.

WHAT ARE POTENTIAL BENEFITS?

Rubrics are traditionally thought of as tools to promote consistent, fair evaluation and grading, and they certainly can have that effect (Hafner and Hafner, 2003; Docktor et al., 2016). They also, however, have additional benefits for students' learning. By specifying the targets that students' work should meet, they clarify the goals of instruction for both students and the instructor (Allen and Tanner, 2006). Ambrose and colleagues identify goal-directed practice as one of the critical elements in learning (Ambrose et al., 2010). Much like learning objectives can help students sharpen their focus and aim their efforts toward the kinds of work their instructors want them to be able to accomplish, rubrics can help students understand what exemplary work looks like and thus give them a target to which to aim (Andrade, 2001; Jonsson and Svingby, 2007). And rubrics have the potential to do more than to clarify the goals of instruction: If students use them well, rubrics can help them be metacognitive, monitoring their work and their thinking to see if it's accomplishing the goals of the assignment and adjusting if it is not.

Students tend to like and value rubrics (Reddy and Andrade, 2010). Clabough and Clabough (2016) investigated student ratings of three instructional methods to promote scientific writing skills in an introductory undergraduate neuroscience class: using a rubric individually as they wished, using a rubric and consulting with a biology subject tutor, or self-grading paper components with the rubric. The students who used the rubric reported more confidence in their writing skills by the end of the semester and rated this use or the use of an example paper as the most effective way to teach scientific writing.

Finally, rubrics have the potential to make grading more efficient, particularly when integrated into a learning management system (e.g., D2L, Canvas, or Moodle) that can automatically convert marks on an electronic rubric to grades in a gradebook and electronic feedback to students. While this may not be the primary reason for adopting rubrics and may not be the only way that instructors choose to use them, it has real benefit for some assignments.

PRACTICAL RECOMMENDATIONS

1. *Choose the type of rubric that fits your assignment, your students, and your grading style.* As described above, holistic, analytic, and single-point rubrics each have their strengths, and the key is to choose the one that is the best fit. Holistic rubrics are often the best choice for low-stakes assignments, especially if they will be repeated. Analytic rubrics provide a level of detail that can be useful for larger assignments, particularly if you want to alert your students to the characteristics that differentiate between "excellent" and "good" performance levels. Single-point rubrics don't detail the qualities of "excellent" work, which may leave space for more creativity and more effort from students, but which may also produce more examples of work that is acceptable but not exemplary. In addition to considering the benefits and limitations of each type of rubric with regard to the assignment, it's also important to consider

which will work best for you. Which will produce grades that you will be comfortable sharing with students? Which will best facilitate your grading? And finally, it's also important to consider how your students will interact with the rubric. Are they more likely to be able to use one of these rubric types to shape and monitor their own work? Are they more likely to understand grades generated with one type of rubric? There may be no perfect answer, but collectively these questions can help you choose a rubric that will work for your context.

2. *Design your rubric to reflect your learning goals and the qualities you want to see in your students' work.* One of the benefits of writing a rubric is that it helps us think through what characterizes excellent performance on a given assignment, converting it from "I know it when I see it" to a description that can help our instruction as well as our students' efforts. A great start to writing a rubric is to describe, in writing, excellent work on the assignment you have in mind. What does that excellent work look like? What are its components? What is done well that makes the work outstanding? It can be helpful to look at examples from previous students or public examples, such as papers or posters, to stimulate your thinking and help you articulate the moves that make the work compelling. Inevitably, as we consider excellent work, we begin to think about work that does not reach that mark, and we can begin to describe work that is fair and work that is unacceptable, noting what is missing or done less effectively that reduces the quality. Common student errors or omissions are a particularly valuable element to include here.

For holistic rubrics, you will simply refine these descriptions until they are sufficiently informative that they are useful for you in grading and your students in constructing the work. If you are writing an analytic or single-point rubric, however, you can mine your initial descriptions for categories, using them to identify the criteria that will comprise the rows of your rubric. Many faculty choose to list the components of the assignment; for example, a lab report might have an abstract, an introduction, a results section, and a discussion. Other faculty members instead describe essential elements, such as "use of evidence" and "description of alternative interpretations." After breaking out the categories, you can then think intentionally about the qualities that differentiate excellent, acceptable, and unacceptable work in each. As you write these descriptions, it's important to be clear and specific. The rubric has several purposes, from making grading more consistent and transparent to helping students produce better work, and they all rely on descriptions that are clear enough to be interpreted reliably. For example, the rubric provided in Table 13.4, used to evaluate a student leading discussion of primary research article in class, clearly differentiates between excellent and good engagement of students in interpreting experiments.

TABLE 13.4 Example Rubric for Student-Led Discussion. Rubric Demonstrates Criteria With Different Weights and Use of Point Ranges

Providing background and context for question	**9–10 points** Provided clear and useful background and context.	**5–8 points** Provided some background, but reason for question somewhat unclear.	**1–4 points** Background unclear and confusing.	**0 points** Did not attempt to provide background for paper.
Engaging students in interpretation of key experiments	**14–15 points** Engaged multiple students in interpreting experiments. Asked questions to extend discussion (e.g., are there different interpretations possible? What were the key controls in this experiment?).	**9–13 points** Provided most interpretation rather than promoting discussion and/or allowed only a few students to interpret experiments and/or did not ask follow-up questions.	**3–8 points** Provided all interpretation, only asking students follow-up questions.	**0–2 points** Did not engage students in interpretation; was dismissive of students' interpretation.
Providing summary and context for results	**9–10 points** Provided clear and useful summary and context for new results.	**5–8 points** Summary or context unclear.	**1–4 points** Both summary and context unclear.	**0 points** No attempt to provide summary or context for results.
Visual aids	**9–10 points** Clear visual aids with useful introductory and summary elements and key elements of figures highlighted.	**5–8 points** Clear visual aids, but lacking introductory or summary elements or highlights of key elements in figures.	**1–4 points** Visual aids unclear or lacking several of the elements noted at left.	**0 points** No visual aids provided.
Managed time effectively	**9–10 points** Effectively managed time during discussion, allotting time for key figures and saving time for summary.	**0–8 points** Less effective time management (e.g., no time for summary; key figure skipped for time reasons; etc.).		

After you have gotten a good start on a rubric by describing the work you would like to see, there are a few key questions to consider.

- *If you use an analytic or single-point rubric, how many categories should you have?* There is, of course, no single answer to this. Generally, it's helpful to have categories that help you and your students consider the important elements of their work, such as clarity of thesis, use of evidence, usefulness of graphics, etc. There should not, however, be so many categories that the rubric makes you throw up your hands in defeat. The rubric should clarify and help you achieve your teaching and assessment goals. It should not be so detailed that the cognitive load of interpreting it takes away from the real goal of doing (for the student) or evaluating (for you) the student's work.

- *In an analytic or single-point rubric, should each of the categories contribute equally to the grade?* As you design your rubric, you should consider which elements of your students' work are most important. For example, you may want your students to make good esthetic choices and to use standard citation norms on posters they generate, but your primary goals probably deal with framing a research question, accurately representing results, and presenting valid conclusions. Giving each of those categories equal value therefore misrepresents your learning goals, suggesting that making an attractive poster is just as important as reporting and interpreting research results. To produce a rubric that signals to students what matters most and that helps you represent those values in your grades, it's important to assign appropriate value to the grading categories. The rubric shown in Table 13.4 focuses on student-led discussions of primary research articles. One category is weighted more heavily than the others, emphasizing the importance of this component of leading discussion.

- *If you use a holistic or analytic rubric, how many performance levels should you describe?* In brief, you should only have as many performance levels as you can realistically differentiate. For many of us, that means there may be three or four, such as excellent, very good/good, acceptable/fair, unacceptable. As with the number of categories, it's important to keep in mind that the goal of the rubric is to clarify expectations and to promote excellent student work. Performance descriptions that highlight—and therefore help students avoid—common errors can therefore be particularly useful. Having too many levels can be counterproductive, either by being overwhelming or by leading to a box-checking mentality that does not lead students to produce their best work.

- *Can a rubric have a range of points within a performance level?* Many rubrics assign a single point value to each performance level; for example, excellent work may receive a 4, good work a 3, etc. Many times, however, a performance level may represent a range of work; "good" work can vary in quality. One way to represent that variation without adding additional performance levels is to use a range of points within a performance level. For example, for a given category on a rubric, excellent work may earn 10 points; good work 8−9 points; fair work 6−7 points. This type of gradation can help represent the range that we know exists within our students' work and can help prevent a box-checking mentality. The rubric shown in Table 13.4 illustrates how point ranges can be used in an analytic rubric.

3. *Consider cocreating a rubric with your students.* One way to ensure a shared understanding of a rubric and to help students think about how to apply it to their work is to cocreate the rubric. To do this, you can spend a few minutes in class talking to students about rubric design and about the goals of your assignment before asking students to brainstorm important categories and performance descriptions for the rubric in small groups. After groups report out, you can take their collected ideas and construct a draft rubric. Review and finalization of the rubric produces a tool that can be valuable for shaping and grading student work. Table 13.5 illustrates a rubric to evaluate class participation that was cocreated with students in a 300-level Biology of Cancer class. The rubric has three categories: attentiveness and participation during lectures; peer review of colleagues' work out of class; and participation during student-led discussions. Students helped develop the specific descriptions of performance levels in each category, and the instructor determined the point range associated with each.

4. *Discuss rubric use with students.* Rubrics have several potential instructional uses that complement their use as evaluative tools. It can be useful to review a rubric when an assignment is made and to talk to your students about using it to plan their work. In addition, students can use the rubric for self-assessment prior to submission, a practice that can be particularly valuable if it is completed several days before the project is due and submitted along with the final work. Alternatively, instructors can set up opportunities for students to use the rubric for peer review of student drafts. Critically reading the rubric as it applies to their colleagues' work can help students see what the rubric descriptions look like in practice, enabling them to apply it more effectively to revisions of their own work.

TABLE 13.3 Example Rubric for Class Participation. Rubric was Collaboratively Developed With Students

	Excellent	Good	Fair
Attentiveness and participation during lecture	All of the following must be true: • No more than three absences. • Consistently visibly attentive. • Not texting, emailing, or using social media in class. • Asking questions and/or making contributions.	Can include the following: • More than three but fewer than six absences. • Inconsistent but usually good visual attentiveness. • Texting, emailing, or using social media a few times in class. • Very few or no questions or contributions.	Can include the following: • Six or more absences. • Inconsistent attentiveness. • Texting, emailing, or using social media several times (>3) in class. • No questions or contributions.
	33–35 points	29–32 points	25–28 points
Peer review of colleagues' work	Thoughtful feedback provided to two colleagues for each of the first four synthesis map submissions. "Thoughtful" is demonstrated by some detail: reference to specific points within the map and what is effective or could be improved at those points.	Feedback provided to two colleagues for each of the first four synthesis map submissions, but with insufficient detail to be helpful on one, two, or three cases.	Feedback not provided one or two times Or Feedback provided each time, but typically with insufficient detail to be helpful.
	20 points	17–19 points	13–16 points
Contribution to student-led paper discussions	At least three active contributions to paper discussions during the semester, taking the form of substantive questions or substantive answers to discussion leaders' questions. (Trivial questions or answers that demonstrate lack of preparation don't count.)	One or two active contributions to paper discussions during the semester, taking the form of substantive questions or substantive answers to discussion leaders' questions. (Trivial questions or answers that demonstrate lack of preparation don't count.)	One active contribution to paper discussions during the semester, taking the form of substantive questions or substantive answers to discussion leaders' questions. (Trivial questions or answers that demonstrate lack of preparation don't count.)
	Consistent visual attentiveness.	Consistent visual attentiveness.	Inconsistent visual attentiveness.
	20 points	17–19 points	13–16 points

CONCLUSION

Rubrics provide a valuable tool to help science faculty be fair and consistent while evaluating open-ended work. They also have the potential to be a versatile instructional tool, helping students understand the goals of an assignment and giving them a means to monitor their work. Rubrics can have different levels of detail, from holistic rubrics that describe different performance levels to single-point rubrics that describe acceptable performance for multiple criteria to analytic rubrics that provide detail on both criteria and performance levels, making them adaptable to different types of assignments, feedback styles, and student populations. Making design choices that are a good fit for your context, perhaps in collaboration with your students, and intentionally incorporating rubric use into your courses can help you maximize the benefits of these tools.

REFERENCES

Allen, D., Tanner, K., 2006. Rubrics: tools for making learning goals and evaluation criteria explicit for both teachers and learners. CBE Life Sci. Educ. 5, 197–203.

Ambrose, S.A., Bridges, M.W., DiPietro, M., Lovett, M.C., Norman, M.K., 2010. How Learning Works: Seven Research-Based Principles for Smart Teaching. Jossey-Bass, San Francisco, CA.

Andrade, H.G., 2001. The effects of instructional rubrics on learning to write. Curr. Issues Educ. 4. Retrieved from: https://cie.asu.edu/ojs/index.php/cieatasu/article/view/1630.

Clabough, E.B.D., Clabough, S.W., 2016. Using rubrics as a scientific writing instructional method in early stage undergraduate neuroscience study. J. Undergrad Neurosci Educ. 15, A85–A93.

Docktor, J.L., Dornfeld, J., Froderman, E., Heller, K., Hsu, Leonardo, Jackson, K.A., et al., 2016. Assessing student written problem solutions: a problem-solving rubric with application to introductory physics. Phys. Rev. Phys. Educ. Res. 12, p010130.

Hafner, J.C., Hafner, P.M., 2003. Quantitative analysis of the rubric as an assessment tool: an empirical study of student peer-group rating. Int. J. Sci. Educ. 25, 1509–1528.

Johnson, A., Svingby, G., 2007. The use of scoring rubrics: reliability, validity and educational consequences. Educ. Res. Rev. 2, 130–144.

Reddy, Y.M., Andrade, H., 2010. A review of rubric use in higher education. Assessm. Eval. High. Educ. 35, 435–448.

Chapter 14

Conclusion

Teaching can be one of the most rewarding elements of a scientist's career, providing a chance to share our fascination with the natural world and our ways of understanding it with a new generation. It also provides ongoing challenges as we deal with new courses, new discoveries, new students, and new technologies. The goal of this book is to help you identify enduring evidence-based approaches that can help shape your courses and your teaching practice to be adaptable, effective, and even (dare we say) relatively efficient.

You may choose to start at the beginning, with course design. You may choose to begin by intentionally incorporating more practices to make your classroom inclusive, or by building in informal group work, or by using rubrics. The beauty of effective teaching practices is that they have multiple benefits and synergies with each other. For example, incorporating active learning approaches helps all of your students learn more and reduces disparities among groups of students. Incorporating retrieval practice into your class makes your students more metacognitive, as does designing a course, class session, or assignment around articulated learning goals. Giving students some autonomy in their assignments increases student interest and makes your class more inclusive. The list can go on.

Thus there is no right or wrong place to begin, and I encourage you to start with the piece that feels most promising to you as an instructor. I also encourage you to be persistent and to think of the principles that underlie any technique or approach you adopt. If it doesn't go exactly as you wish the first time, think about why it didn't, make some tweaks, and try again. Ask a colleague in your department or your center for teaching and learning for advice. Or ask your students. One of the aims and methods of effective teaching is to have you and your students be on the same team, working toward common goals. Telling them why you are doing something and asking them for their feedback can be a powerful way to make that happen and to help you hone your teaching practice.

So dive in! Choose the piece that will help you and your students, adapt as you and they need, and have fun.

Science Teaching Essentials. DOI: https://doi.org/10.1016/B978-0-12-814702-3.00024-X

Index

Note: Page numbers followed by "*f*" and "*t*" refer to figures and tables, respectively.

P

Pause procedure, 67
  lecturing, 112, 116–117
Peerceptiv, 41–42
Peer evaluation, 81
Peer instruction, 68, 74*f*, 122–123
Peer-led team learning, 70
Peer review, 41–42
Peer-to-peer interaction and support, 9
Performance, 5–6
Perseverance, 10–11
Personal electronic devices, 51
Physics, 16
  course-based research in, 154
  core ideas, 19*t*
Piaget, Jean, 63, 123
Planning, 88–89
Poll Everywhere, 68
Preclass work, 127*t*
Prior knowledge activation, and lecturing,
  112–115
Problem-based learning (PBL), 70, 73–74
Problem-solving exercises, for metacognitive
  skills, 90–91
Procedural knowledge, 32, 88
Process-oriented guided inquiry learning
  (POGIL), 70, 73–74, 80
Process scaffolding, 39–40, 40*t*
Productive feedback, 122
Publication bias, 64–65
Public speaking skills, 111–112

Q

Questions into the videos, integrating, 138

R

Recall lecture information, 117–118
Recognition, 5–6
Redundancy elimination, in educational
  videos, 139
Reflection, 35
  questions, enhanced answer keys with, 90
Reliability of exam construction, 164–165
Research Corporation for Science
  Advancement Cottrell Scholars, 15–20
Retrieval practice, 67, 95
  lecturing, 111, 117–118
  and long-term retention, 96–97
Rote memory, and test-enhanced learning,
  99–100

Rubrics, 41–42, 175
  analytical, 175–177, 176*t*, 181
  for class participation, 182, 183*t*
  constructed-response questions, 171*t*
  holistic, 175, 176*t*, 177, 181
  potential benefits of, 177–178
  practical recommendations for, 178–183,
    180*t*
  single-point, 176–177, 177*t*, 181

S

Scaffolding, 37
  cognitive skills, 40–41, 40*t*
  evaluation, 40*t*, 41–42
  exams, 44–45
  process, 39–40, 40*t*
Science communication, 145*b*
Science Education Alliance–Phage Hunters
    Advancing Genomics and
    Evolutionary Science
    (SEA–PHAGES), 150
Science identity, 5–7, 7*f*
Scientific practices, 17*t*
  course-based undergraduate research
    experiences, 147
Scientific Teaching, 61–62
Scientists of color, 9
Segmenting, 136–137
Selected response items, 167–168, 167*f*
Self-assessment checklists, 91–92
Self-directed learners, 85
Self-efficacy, 6–7, 10–11, 35, 36*f*
  promotion of, 37
Self-esteem, 75
Self-regulated learning, 10–11
Self-regulatory learning, 90
Self-select groups, 79
Self-system model for classroom motivation,
  18, 36*f*
Self-system model of supportive classroom
    environment, 6–7
Sense of belonging, 6–10, 35
Short-answer test, 96–99
Signaling, 137
Single-point rubrics, 176–177, 177*t*, 181
Skype, 57
Slater, Stephanie, 40–41
Slater, Tim, 40–41
Small-group learning, 75–76, 79
Small World Initiative, 149–150
Social-belonging intervention, 10

Printed in the United States
By Bookmasters